Philipp Krüger

Optimisation of hysteretic losses in high-temperature superconducting wires

Karlsruher Schriftenreihe zur Supraleitung   Band 015

HERAUSGEBER

Prof. Dr.-Ing. M. Noe
Prof. Dr. rer. nat. M. Siegel

Eine Übersicht über alle bisher in dieser Schriftenreihe
erschienene Bände finden Sie am Ende des Buches.

Optimisation of hysteretic losses in high-temperature superconducting wires

by
Philipp Krüger

Dissertation, Karlsruher Institut für Technologie (KIT)
Fakultät für Elektrotechnik und Informationstechnik, 2013
Hauptreferent: Prof. Dr.-Ing. Mathias Noe
Korreferent: Prof. Frédéric Sirois, B.Ing., Ph.D.
Betreuer: Dr. Francesco Grilli

Impressum

 Scientific
Publishing

Karlsruher Institut für Technologie (KIT)
KIT Scientific Publishing
Straße am Forum 2
D-76131 Karlsruhe

KIT Scientific Publishing is a registered trademark of Karlsruhe
Institute of Technology. Reprint using the book cover is not allowed.

www.ksp.kit.edu

 *This document – excluding the cover – is licensed under the
Creative Commons Attribution-Share Alike 3.0 DE License
(CC BY-SA 3.0 DE): http://creativecommons.org/licenses/by-sa/3.0/de/*

 *The cover page is licensed under the Creative Commons
Attribution-No Derivatives 3.0 DE License (CC BY-ND 3.0 DE):
http://creativecommons.org/licenses/by-nd/3.0/de/*

Print on Demand 2014

ISSN 1869-1765
ISBN 978-3-7315-0185-5
DOI: 10.5445/KSP/1000039387

Optimisation of hysteretic losses in high-temperature superconducting wires

Zur Erlangung des akademischen Grades eines

DOKTOR-INGENIEURS

von der Fakultät für Elektrotechnik und Informationstechnik

des Karlsruher Instituts für Technologie

genehmigte

DISSERTATION

von

Dipl.-Ing. Philipp Andreas Christoph Krüger

geb. in: Leoben

Tag der mündlichen Prüfung: 14. Februar 2014
Hauptreferent: Prof. Dr.-Ing. Mathias Noe
Korreferent: Prof. Frédéric Sirois, B.Ing., Ph.D.
Betreuer: Dr. Francesco Grilli

Acknowledgement

This dissertation, while of course being original and of my own doing, often relied on the help of others. First and foremost, I would like to thank my advisor Dr. Francesco Grilli for his unremitting support. The possibility to always come asking when being stuck and to be able to discuss scientific and other matters anytime were a priceless help. The freedom to pursue my own scientific interests and to conduct studies and research abroad are exceptional as were the possibilities to visit any conference we felt might benefit my work.

Prof. Mathias Noe, my doctoral thesis supervisor, besides making sure the work proceeded steadily, provided support for extraordinary experiences that made my stay at the Institute for Technical Physics so memorable, be it master courses at the ESADE business school in Barcelona or the seminars in Austria and in the Black Forest. He also helped raise the quality of this work considerably by providing ideas and feedback and ensured the claim for excellence.

The collaboration with Dr. Victor Manuel Rodriguez Zermeño was particularly help- and fruitful and his advice invaluable.

The other post-doctoral researchers in my group, Dr. Eduard Demencik and Dr. Michal Vojenciak, both helped me understand the theory and provided much needed help while conducting experiments. So did Andrea Kling, who also contributed greatly to the success of the experiments. Dr. Christian Barth not only helped me set up measurements but also was a brother in arms during most of my time in Karlsruhe and I enjoyed our time together a lot. Dr. Alexandra Jung and Dr. Rainer Nast both helped me out with measurements and advice.

During my stay at École Polytechnique in Montréal, I had the chance to work with Prof. Frédéric Sirois who also took the time to serve as my second thesis supervisor and I am thankful for both. I also enjoyed working with Dr. Michael J. Cheadle, Franco Julio Mangiarotti, Dr. Joseph Minervini and Dr. Makoto Takayasu tremendously during my time at the Massachusetts Institute of Technology. For involving themselves in my dissertation I would like to thank Prof. Dr.-Ing. Sören Hohmann and Prof. Dr.-Ing. Martin Doppelbauer.

The friendly atmosphere at our institute is due to the great company of friends and colleagues such as Dr. Nadja Bagrets, Christoph Bayer, Dr. André Berger, Florian

Erb, Enno Heits, Sebastian Hellmann, Oliver Näckel, Dr. Enrico Rizzo, Dr. Christian Schacherer, Anne-Kathrin Weber, Dr. Klaus-Peter Weiß, Teresa Schwarz, Dr. Michael Schwarz and Stefan Zimmermann.

I would not have been able to make it this far without my family, for which I shall ever be indebted.

gaudeamus igitur
iuvenes dum sumus

Zusammenfassung

Um die im Titel der Dissertation erwähnten Optimierungen der Kabel aus Hochtemperatursupraleitern durchführen zu können, wird ein numerisches Modell vorgestellt, welches in der Lage ist, das physikalische Verhalten von Supraleitern und Ferromagneten zu simulieren. Dieses wird mittels Magnetisierungs- und Transportstrommessungen an Supraleiter/Ferromagnet-Heterostrukturen verifiziert. Die angesprochenen Heterostrukturen bestehen aus einem sogenannten Bandleiter (Hochtemperatursupraleiter zweiter Generation) mit ferromagnetischen Schilden, die die Hystereseverluste durch Herumleiten des magnetischen Flusses um den Supraleiter herabsetzen sollen. Neben den Bandleitern werden verschiedene Spulengeometrien untersucht. Auch untersucht wird das sogenannte Twisted Stacked Tape Conductor Kabel, welches für Hochstromanwendungen, beispielsweise in der Fusionstechnologie, entwickelt wird. Hierfür wird das zweidimensionale numerische Modell um die dritte Dimension erweitert und außerdem werden Möglichkeiten untersucht, durch Rand- oder Zwangsbedingungen die Kontaktwiderstände zu berücksichtigen. Dies ist notwendig, wenn die Translationssymmetrie des Kabels beibehalten werden soll und die Kontaktwiderstände nicht als eigenständige geometrische Objekte in Erscheinung treten sollen. Bei der Optimierung werden verschiedene Geometrien vorgestellt, in welchen Hystereseverluste durch ferromagnetische Schilde zum Teil um eine Größenordnung herabgesetzt werden können.

Contents

List of Figures

1 Introduction

Increasing industrialisation and automation all over the world in combination with increasing scarcity of conventional resources on the one hand and political decisions to abandon existing power generation technology on the other hand are leading to rapid evolution in the energy sector [SSA⁺07]. The shift from conventional sources of power generation is being made to sustainable and renewable energy sources such as wind, solar and tidal forces [Pre11]. This development in combination with breakthrough discoveries in material science having lead to high temperature superconducting materials in the last two and a half decades promote innovative applications, especially in the energy but also in the transportation sector [BM86, NNM⁺01, LGFP01, Mal12]. Increasingly localised power generation requires efficient energy distribution and the temporal variation of power generation due to unsteadily available wind and solar power entails the need for energy storage. New technologies and innovative solutions are required to overcome these challenges and superconductors have great potential [NEM⁺13].

Superconducting applications differ considerably in their state of implementation. Some are merely exotic concepts, such as the all electric air-plane for which rotating machinery, power electronics and linear motors are required [MBSL07, MCT⁺09, LMN⁺09]. Others like superconducting motors for naval propulsion have passed the demonstrator stage and are tested on regular duty [NFK⁺10, UAY⁺10, MSU⁺13]. Another example of superconducting rotating machinery being investigated at present are superconducting wind turbine generators [LM07, AMS⁺09, AMS⁺10, AJS⁺11]. Superconducting wind turbines could significantly reduce the overall weight of the nacelle thereby considerably reducing mechanical stress on the support structure and thus reducing the investment cost while at the same time reaching a higher power than would be possible using conventional technology [AMS⁺10, SGKW11, FOS⁺11, TSO12]. Superconducting fault current limiters have been around for a long time [GF78, BP83, Row95, NS07] and demonstrator as well as commercial appliances are available [DKH⁺10, HEB⁺12, KSC⁺12]; transformers have not been commercialised but several demonstrator projects exist [KHH⁺02, IHO⁺09, BTI⁺05, KHF⁺06, WWS⁺09, WZH⁺08, GSB⁺11, KKK⁺11, TOO⁺11], for superconducting magnetic energy storage systems [JGK⁺02, NHK⁺04, NHS⁺04, KSC⁺06, OCA⁺06, KNN⁺09, XWD⁺08, KKL⁺06, TBD⁺07, KLL⁺09] and for superconducting high current, medium voltage cables [KKT$^+$99, KWN$^+$01, Mas02, MKY$^+$02, HHJ$^+$04, XHB$^+$04, KJL$^+$05, SCK$^+$06, CBK$^+$06, Mal06, MSB$^+$07, DSJ$^+$07, TYM$^+$07, SHL$^+$07, Tsu08, YAI$^+$09, MSBW09, MFL$^+$09, MKH$^+$09, SVF$^+$10, SMN$^+$12] which are being tested in commercial power grids. Persistent mode very high field coil systems are being developed that use superconducting coated conductors [MKO$^+$12, MLW$^+$12, MBJC13, WMV$^+$13]. Superconductors are also interesting as bulk material for high magnetic field permanent magnets [HM04, CLY$^+$11, XLC$^+$12, NSF12, WFDR$^+$12].

The significant increase in critical temperature from low temperature superconductors to high temperature superconductors results in a considerable reduction of the cooling cost [LGFP01, GFH$^+$08]. Instead of expensive liquid Helium, liquid Nitrogen may be used as the cryogenic medium for the high temperature superconductors, e.g. of the oxocuprate class due to their high critical temperature [BM86]. Conduction cooling is also possible [YHP$^+$07, DYZ$^+$06, TOK$^+$05].

Oxocuprate superconductors are exceedingly interesting technologically not least due to their extremely high current density resulting in minuscule material volume requirements [GNB$^+$96]. Two possibilities exist at present to manufacture conductors using these materials: the first generation was produced via the powder-in-tube technique [HTT89]. The second generation of these ceramic superconductors is deposited as a thin film layer on top of various metallic [GNB$^+$96, GNK$^+$97, BSK99, GAF$^+$02] and ceramic substrates. Possible ceramic substrates used as wafers for superconducting circuitry include corundum varieties such as sapphire (α-Al$_2$O$_3$) [MJBM88, TWP89]. They are of little importance for energy applications considered subsequently and will not be expatiated upon any further. Instead the metallic coated conductor tapes are explored which are presently the mainly manufactured oxocuprate superconductors [MFR$^+$08].

The particular geometry of the superconducting layer of the coated conductors which is about 1 cm wide and only 1 μm thick and thus has high aspect ratios up to or even above 10 000 entails a peculiar behaviour that is not intuitively predictable, especially so since superconducting materials per se behave quite differently from regular conductors. Several analytical models exist that are very successful in describing the behaviour of coated conductor takes under selected boundary conditions [Bea64, Bea62, BD62, Nor70, BI93, Bra96, Maw96, MC01, Kaw01, Maw08]. Despite their idealisations, restrictions regarding boundary conditions, material properties and geometry they yield qualitatively excellent results and describe the superconducting behaviour well. Unless symmetries are exploited, modelling multiple superconducting domains is taxing however and not all arbitrarily complex geometries can be considered. The case becomes even more difficult if additional material classes such as ferromagnetic domains are present [Maw08]. The analytic models will be used in order to verify the numerical models in Cap. 5, Cap. 6 and Cap. 7.

Hysteretic losses are an important heat source in superconductors when AC electric currents and magnetic fields are applied. A single 10 mm wide coated conductor dissipates around $0.5\,\mathrm{W\,m^{-1}}$ at I_a/I_c [Nor70, GPS$^+$14]. Elevated energy dissipation results in a higher heat load on the refrigeration units. Even an ideal cryocooler is limited by the Carnot efficiency. This results in the refrigeration unit requiring at least 2.9 W theoretically for every Watt dissipated at 77 K. The ratio worsens the lower the temperature is: at 30 K, 9 W are required and at 4.2 K, 70 W are required [KC07, p. 1195]. Realistic estimates even for 77 K put the cooling penalty factor at 15 and to over 1000 for 4.2 K [RLR$^+$98, GPS$^+$14].

Though some applications are already capable of competing with conventional technology [YKT$^+$14] most superconducting applications still suffer from high investment and operating costs which reduces their economic appeal [MH12]. The investment costs are higher than those of conventional machines because on the one hand the superconductors are still too expensive to be economically competitive and on the other hand there is the additional cryogenic system to be considered, which is required for keeping the superconductors operational. On the operating cost side the additional cooling costs have to be taken into account, even if the superconductors result in lower operating costs than a comparable conventional machine. The cost of superconducting material is expected to become more accommodating in the future [MH12, Haz13].

Besides capitalising upon the effects of scale and technological advance in materials science, there is a material independent approach to reducing both investment as well as operational costs: reducing hysteretic losses and optimising the appliances by geometry and material combination optimisation. The reduced requirements on the cryogenic system due to decreased thermal load lead to lower investments in the refrigeration unit and also to reduced operating costs because of the reduced energy requirement of the cooling system. As will become apparent in this thesis, the reduction in hysteretic losses achievable by design optimisations is substantial.

In order to optimise superconducting applications, appropriate tools have to be available to allow researchers and developers to analyse and design machines. Numerical modelling provides fast and cheap methods towards that purpose.

At the outset of the thesis was a survey of the possibilities to improve the hysteretic loss of superconducting applications and possibly provide tools to ease the design process.

Several numerical modelling approaches exist, the $T - \Omega$- [HSM$^+$91, AMBM98], the A- [NTFA89, Seb94, Pri97, BMDH99, CFD99, CFD99, HYT$^+$00, ST10], the electrostatic-magnetostatic analogue [GRKN09], the magnetic vector potential critical state [Cam07, Cam09], or the adaptive resistivity implementation [FFG10]. The time-dependent H-model was used extensively [BV83, KTK$^+$01, KHY$^+$03, PMC$^+$03, HCC06, BGM07, NAW09] but no three dimensional simulations of actual cables have been shown so far.

The goal of this thesis is to demonstrate a comprehensive implementation of a numerical model capable of simulating superconducting as well as ferromagnetic domains in two and three dimensions. Besides the ability to simulate magnetic field and electric current distributions and calculate hysteretic losses, the model should be modular in order to allow additional functionality to be implemented. The model is extensively verified using analytical formulations and experimental data. Additionally, a new boundary condition for finite element modelling is developed in order to be able to simulate contact resistances.

The numerical model is used to analyse various coated conductor geometries like isolated tapes, bifilar and pancake coils with and without ferromagnetic shielding in order to verify the possibility of reducing the hysteretic losses. A three dimensional model is used to model the behaviour of the Twisted Stacked Tape cable. Apart from the theoretical work, the concept of ferromagnetic shielding was also investigated experimentally. A method for applying ferromagnetic material on superconducting tapes was developed and samples were measured in applied magnetic background field and electric transport current.

The structure of this thesis is as follows: the fundamentals of superconductivity and ferromagnetism are explained in Cap. 2. Then, the concepts of analytical and numerical modelling of superconductors are explored in Cap. 3. The experimental set-ups are detailed in Cap. 4. The physical behaviour of coated conductors is discussed in Cap. 5 and that of various assemblies with and without ferromagnetic shielding in Cap. 6. Lastly, the Twisted Stacked Tape Cable is investigated with analytic and, more importantly, a three-dimensional numerical model in Cap. 7.

2 Fundamentals

Superconductivity is a physical phenomenon observable in a number of materials, among them pure metals, alloys, inorganic compounds – most prominent among those: ceramics – and even organic compounds (see Cap. 2.7). The most eminent property of superconducting materials is their vanishing electric resistance below the critical temperature T_c (see Fig. 2.1). This effect was first recorded by Heike Kammerling Onnes in 1911 [Onn12]. The extremely low energy dissipation (due to extremely low resistance) and high energy density in applications are among the main reasons for interest in these fascinating materials. In addition to conducting electric currents entirely without dissipation in certain cases, superconductors differ from a theoretical perfect conductor in that they expel all magnetic flux from their volume; this behaviour is called the Meißner-Ochsenfeld-effect [MO33].

There are two types of superconductors: conventional Type-I superconductors which show a perfect Meißner-Ochsenfeld-effect and Type-II, also called technical superconductors. All pure elemental superconductors with the exception of Hafnium, Niobium, Tantalum, Titanium, Thorium, Vanadium, Uranium and Zirconium are Type-I superconductors [WC52]. Due to thermodynamical reasons, in Type-II superconductors a

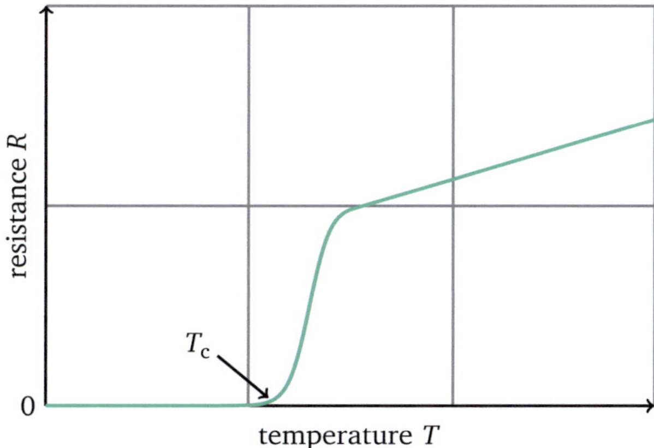

Figure 2.1: In superconducting materials, the electric resistivity vanishes below a characteristic temperature. T_c is determined by measuring the DC resistivity of a sample. However, in order to prove superconducting behaviour, a jump in resistivity is not sufficient. The Meißner-Ochsenfeld-effect has to be shown [Tin96, pp. 2 & 3].

state of bulk superconductivity with locally confined volumes of magnetic flux is energetically more convenient (see Cap. 2.1). Since by definition, regions with magnetic flux cannot be superconducting because they do not fulfil the prerequisite of showing the Meißner-Ochsenfeld-effect, those locally confined volumes of magnetic flux are normal conducting [MU97, pp. 101 ff.]. One single such volume is called a fluxon or, more commonly, a flux vortex. Vortices are of importance when considering losses in superconductors because in normal conducting volumes conventional loss mechanisms are present. In order to facilitate understanding of hysteretic loss mechanisms in superconductors, an introduction to superconductor theory is given below.

Some superconductors are deposited on ferromagnetic material that influences the magnetic field distribution and thus has influence on the behaviour of the superconductor. This influence is investigated and the concept of ferromagnetic shielding is explored. The shielding is supposed to reshape the magnetic field around a superconductor thus rerouteing flux to lower the hysteretic losses occurring in the superconductor. In Cap. 2.8, basic properties of ferromagnetic materials are discussed.

2.1 Theories of Superconductivity

The London-Theory, published in 1935 [LL35], accounts for the already mentioned behaviour of vanishing electric resistivity and the Meißner-Ochsenfeld-effect. It first proposed the London penetration depth λ_L, see Cap. 2.1.1. The phenomenological Ginzburg-Landau-Theory, published in 1950 [GL50], accounts for macroscopic as well as quantum effects. It proposed a second characteristic parameter important in superconductor theory, the coherence length ξ, see Cap. 2.1.2. Without deriving the theories completely, those characteristic parameters will be explained in the following chapters. Lastly, the Bardeen-Cooper-Schrieffer-theory [BCS57] provides a microscopical explanation of superconductivity (which explains classical superconductors).

2.1.1 The London Penetration Depth

The London-theory [LL35] was the first to describe the currents flowing in and the electromagnetic fields generated by superconductors. Magnetic fields were found to cause so called shielding currents to flow in surface regions of the superconductor and generate fields of equal magnitude to the external fields thereby effectively shielding the superconducting volume from magnetic flux entry. Shielding currents only flow in thin surface regions penetrating a characteristic length called the London penetration depth λ_L. Solving the London equations, the characteristic length λ_L over which an external applied magnetic field is suppressed to $1/e$ its initial strength inside the superconductor is found to be [MU97, p. 22]:

$$\lambda_\mathrm{L} = \sqrt{\frac{m}{\mu_0 n_\mathrm{s} q^2}} \tag{2.1}$$

with the charge mass m, the permeability constant μ_0, the charge particle density n_s and the charge q. The charge q was later found to be twice the electron charge e^- [Coo56, Gor59, GMB64].

2.1.2 The Coherence Length

In addition to the already mentioned London penetration depth, the macroscopic Ginzburg-Landau-theory [GL50] (also: Ginzburg-Landau-Abrikosov-Gor'kov-theory) predicts another characteristic length in superconducting materials, the coherence length ξ. The theory is built upon the understanding that the superconducting charge ensemble – the entirety of all superconducting charge carriers – may be described by the complex order parameter field ψ. ξ is the material specific characteristic length inside of which ψ may be assumed constant. It was later discovered that ψ is a measure of the charge particle density $n_\mathrm{s} = |\psi|^2$ [Tin96, p. 9].

In the scope of the microscopic Bardeen-Cooper-Schrieffer-theory [BCS57], the coherence length may be understood as the average distance at which two electrons interact to form a Cooper-pair, the basic superconducting charge carrier. In order for electrons to pair, their Coulomb repulsion has to be overcome. In conventional superconductors, this is brought about through interaction of two electrons via phonons (a weak interaction): supposing an electron moves along a metal's crystal lattice, then the positively charged lattice will be attracted by the negatively charged electron. Due to the high electron velocity v_e and the comparatively high inertia of the lattice resulting in a low Debye frequency of the phonon lattice w_D, the maximum deformation of the lattice will be reached in a distance of $v_\mathrm{e} 2\pi/w_\mathrm{D} \approx 100$ nm where $v_\mathrm{e} \approx 10^6 \mathrm{m\,s^{-1}}$ and $2\pi/w_\mathrm{D} \approx 10^{-13}\mathrm{s}$. This characteristic length is large enough to effectively shield the Coulomb repulsive force, meaning the electrons can interact and couple (compare [MU97, pp. 150-152]).

The mechanism of superconductivity in Type-II superconductors is not yet fully understood and it has been questioned whether it is indeed due to electron-electron interaction. It would seem that the mechanism involving phonons alone, a weak interaction, insufficiently explains the so-called d-pairing observed in a number of high-T_c superconductors, particularly in the technically interesting oxocuprates (see Cap. 2.7.3). The strong interaction required to understand superconductivity in high-T_c superconductors could possibly be connected to mechanisms involving electron deficient holes, spin-wave coupling or excitonic processes. For the scope of this introduction, the knowledge that an interaction exists is sufficient however.

The Ginzburg-Landau-theory introduces the Ginzburg–Landau parameter κ which is the ratio of λ_L and ξ:

$$\kappa = \frac{\lambda_L}{\xi} = \sqrt{\frac{m^2 \beta}{2\mu_0 q^2 \hbar^2}} \tag{2.2}$$

with the Dirac constant $\hbar$ and the empirical Ginzburg-Landau parameter β. The Ginzburg–Landau parameter κ will be useful for characterising superconductors into Type-I and Type-II, see Cap. 2.2.

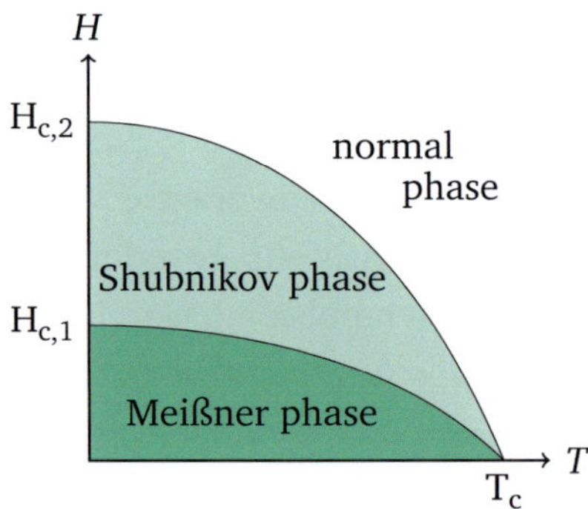

Figure 2.2: In conventional Type-I superconductors only the Meißner-phase exists, bounded by material specific critical parameters H_c and T_c; in Type-II superconductors, the critical magnetic field $H_{c,1}$ below which the perfect Meißner-Ochsenfeld effect is observable is usually negligibly small for technical applications. However, there exists another state: the Shubnikov-phase in which magnetic flux penetrates the superconductor. The Shubnikov-phase is bounded by $H_{c,2}$ and T_c.

2.2 Type-II Superconductors

The Ginzburg-Landau-theory proposes the existence of superconductors in which the interface energy between normal and superconducting domains is positive. Such materials were indeed discovered later on to have magnetic flux entering above a certain threshold magnetic field strength called the lower critical magnetic field $H_{c,1}$. Above this threshold magnetic field, the superconductor enters the Shubnikov phase (see Fig. 2.2). For an overview of the superconducting materials and the most important Type-II superconductors, see Cap. 2.7.

Type-II superconductors have, due to the mechanism of magnetic flux entering their volume, one crucial advantage when compared to Type-I superconductors: they tolerate much higher magnetic fields before losing their superconducting properties completely. Above the lower critical magnetic field $H_{c,1}$, magnetic flux may penetrate the superconductor but the superconducting state is preserved in most of the volume. At the upper critical magnetic field $H_{c,2}$, all bulk superconductivity is lost but this is usually much higher than the thermodynamic critical magnetic field $H_{c,th}$ which marks the loss of superconducting properties in Type-I superconductors (see the plot of the sample magnetisation M versus magnetic field H in Fig. 2.3). The technically more interesting irreversibility field is also much higher, often in the range of multiple tens of Tesla. The magnetic irreversibility field is defined as the magnetic field above which the critical current becomes vanishingly small at a given temperature. For example, the $H_{c,th}$ of all Type-I superconductors is in the range of < 200 mT [Kit96, p. 336] while the most important Type-II superconductors have an upper critical magnetic field $H_{c,2}$ of over 200 T [BK04, pp. 82 & 99]. The barium strontium calcium oxocuprates (henceforth termed BSCCO) only have a magnetic irreversibility field of 200 mT at 77 K but at lower temperatures they can tolerate much higher fields and at liquid Helium temperatures they can be used at field strengths exceeding 10 T [vdLvEtH$^+$01]. NbTi, Nb$_3$Sn and MgB$_2$ show similar magnetic properties, although they only become superconducting at much lower temperatures. The rare earth barium oxocuprates (henceforth termed REBCO) can even tolerate magnetic fields over 30 T [HKC$^+$07].

The penetration of magnetic flux into the superconductor volume is governed by the competing energetic influences of the loss of condensation energy and the conservation of energy saved by not having to displace magnetic flux. Upon flux penetration, condensation energy is lost because electrons that would otherwise tend to form Cooper pairs due to the superconducting state in general being energetically more favourable below a temperature $T < T_c$ fail to do so. On the other hand, since the

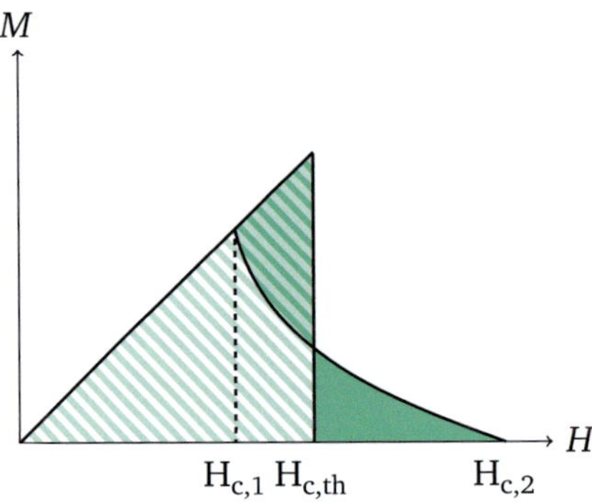

Figure 2.3: The lower critical magnetic field $H_{c,1}$ denotes the field amplitude above which magnetic flux starts to penetrate the superconducting volume resulting in changes to the sample magnetisation M; above the upper critical magnetic field $H_{c,2}$ bulk superconductivity is not observable. The thermodynamic critical magnetic field $H_{c,th}$ is the thermodynamic critical field that in the case of Type-I superconductors marks the onset of the normal conducting state; in the case of Type-II superconductors it is the magnetic field at which the difference of the integrals of the magnetisation curves below and the integral of the magnetisation curve above $H_{c,th}$ equal.

process of displacing magnetic flux from inside the superconductor expends energy and allowing some flux to enter its volume saves a fraction of said energy, a balance arises that depends on the two factors. Consider the interface surface energy σ which is approximately defined by [MU97, pp. 57-60]:

$$\sigma \approx \frac{1}{2}\mu_0 H^2 (\xi - \lambda_L).\tag{2.3}$$

with the magnetic field H. In conventional superconductors $\lambda_L < \xi$ so that $\sigma > 0$ where in Type-II superconductors $\lambda_L > \xi$ so that $\sigma < 0$. Tab. 2.1 shows the classification of superconducting materials into the two possible classes. Type-II superconductors consequently profit energetically from flux entering the volume. At the same time, in order for the superconducting state to persist, some domains need to remain flux free. This leads to the question of what the resistive domains look like which will be discussed in Cap. 2.3.

2.3 Magnetic Flux Quantisation

Since the positive interface energy entails an energy gain when the interface is maximised, it is energetically convenient for the magnetic flux to segment into a

Classification of superconductors	
Type-I	Type-II
$\kappa < 1/\sqrt{2}$	$\kappa > 1/\sqrt{2}$
$\sigma > 0$	$\sigma < 0$

Table 2.1: Using the Ginzburg–Landau parameter κ (see Eq. 2.2) and the interface surface energy σ to classify superconducting materials. When rigorously solving the Ginzburg-Landau theory and not using the approximate formula given in Eq. 2.3, it can be shown that σ is zero if κ is exactly $1/\sqrt{2}$. When κ is larger than $1/\sqrt{2}$, the coherence length ξ is shorter than the London penetration depth λ_L and the material is Type-II and vice versa. Similarly, when σ is smaller than zero, it is energetically more convenient to allow a certain amount of magnetic flux to penetrate the superconductor volume which is characteristic of Type-II superconductors.

multitude of disconnected volumes instead of staying singly connected. This leads to the formation of fluxons, or flux vortices, which usually arrange in distinctive geometric patterns. With cryogenic scanning tunnelling microscopy, it is possible to observe these vortex patterns in situ [SSS+03] (compare Fig. 2.4). Quantum mechanics dictates a lower limit for the amount of flux contained in any such vortex.

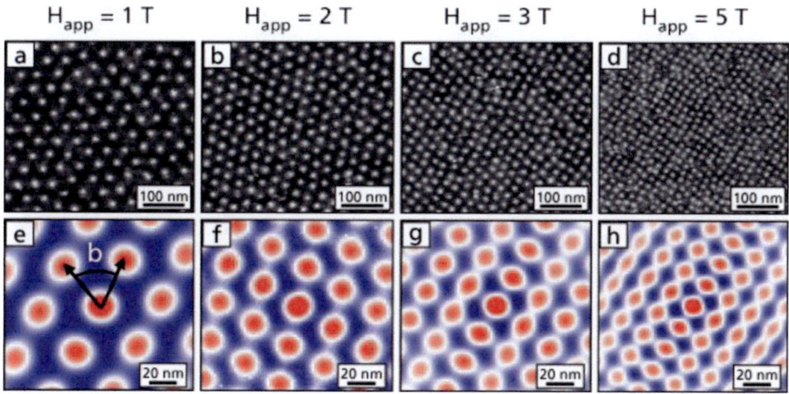

Figure 2.4: (a-d) STM Fermi-level conductance images of the vortex lattice of V_3Si as a function of applied magnetic field at 2.3 K. (e-h) Corresponding auto-correlation images showing the unit cell of the vortex lattice undergoing a hexagonal-to-square symmetry transition. Reprinted figure with kind permission from Sosolik et al., Physical Review B **68**, 140503/1-4, 2003. Copyright 2003 by the American Physical Society [SSS+03].

Consider the Cooper-pair ensemble being described by the complex order parameter field ψ, in this instance interpreted as the macroscopic wave function. After a revolution along an arbitrarily chosen path integral contour C, when returning to the place of origin, ψ has to retain its initial amplitude and phase. Then it is intuitively understandable that magnetic flux Φ is quantised in superconductors since the London theory states that the electric current $\vec{J}$ is a function of the phase of the macroscopic wave function ϕ and the magnetic vector potential $\vec{A}$:

$$\vec{J} = \frac{n_s q \hbar}{m} \nabla \phi - \frac{n_s q^2}{m} \vec{A}. \tag{2.4}$$

Due to the constraint mentioned earlier and because ϕ is defined up to integer multiples of $2 \cdot \pi$, we get:

$$\psi = \sqrt{n_s} e^{i\phi} = \sqrt{n_s} e^{i(\phi + 2\pi n)} \tag{2.5}$$

with the integer n. The total variation of the phase along the closed path integral contour C is thus:

$$\Delta \psi = \oint_C \nabla \phi \cdot d\vec{l} = 2\pi n. \tag{2.6}$$

with the infinitesimal length $\vec{l}$, which, using Eq. 2.4, leads to the condition:

$$\oint_C \vec{J} \cdot d\vec{l} + \frac{n_s q^2}{m} \oint_C \vec{A} \cdot d\vec{l} = 2\pi n \frac{n_s q \hbar}{m}. \tag{2.7}$$

Choosing the path integral contour C far from the surface/vortex core (at a distance much larger than λ_L), the supercurrent $\vec{J}$ vanishes, leading to

$$q \oint_C \vec{A} \cdot d\vec{l} = q\Phi = 2\pi n \hbar. \tag{2.8}$$

The magnetic flux Φ is thus quantised in integer units of the magnetic flux quantum Φ_0 (fluxoid):

$$\Phi_0 = \frac{2\pi \hbar}{q} = \frac{h}{2e^-}. \tag{2.9}$$

where we used $q = 2e^-$ and the Planck constant h. The magnetic flux quantum Φ_0 has also been experimentally measured to be [DN61]:

$$\Phi_0 = 2.067833758(46) \text{ fWb}. \tag{2.10}$$

2.4 Flux Pinning and Critical Current

If isotropic material properties of superconductors in the Shubnikov phase were assumed, there should be a measurable loss due to the vortices moving through the material. The movement stems from the Lorentz force acting on the vortices, displacing them perpendicular to the transport current and the resulting magnetic field (see Fig. 2.5). The losses occurring during the movement of the vortices stem from the normal-conducting electrons in the vortex cores interacting with the crystal lattice [MU97, pp. 132 ff.]. The fact that Type-II superconductors may carry dc electric current without dissipation under certain conditions gives rise to the assumption that this may not be entirely correct. Indeed, it is well known that if the applied magnetic field is not too high and the thermal energy is below a threshold, the vortices will stay fixed in place [BK04, pp. 282 ff.].

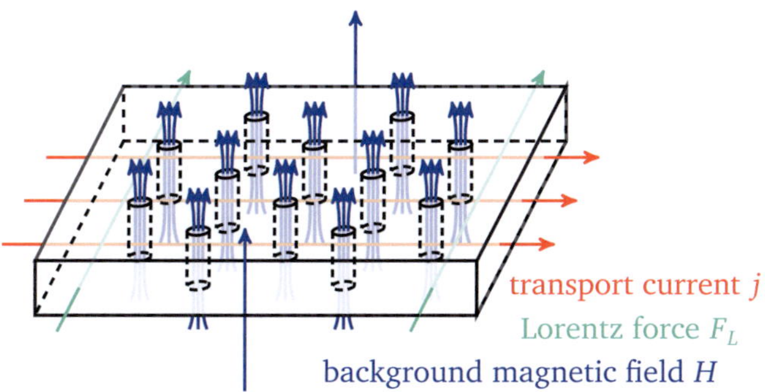

Figure 2.5: A Lorentz force acting on the flux vortices results from the electric transport current and a background magnetic field of depicted direction. The Lorentz force displaces the vortices laterally with respect to the transport current and perpendicular to the magnetic background field. If no active pinning centres are present, the movement is unimpeded and the superconductor is in the so called flux flow regime.

The sites at which vortices stay at a fixed position are called pinning centres and act as potential wells, reducing the total energy of the vortices. There is a range of pinning centres, both intrinsic to the material as well as artificially introduced: zero-dimensional pinning centres such as crystal lattice point defects, one-dimensional pinning centres such as displacement lines, two-dimensional pinning centres such as grain and domain boundaries or surface defects and three-dimensional pinning centres such as artificially introduced precipitations and columnar defect channels introduced by ion or electron bombardment. A material without any pinning centres whatsoever would only be of limited technical use since its critical current would be zero.

With the vortices fixed in place, no dissipation is observable. If the Lorentz force acting on a particular vortex becomes larger than the pinning force however, the vortex will tear free of the pinning centre. It will continue moving through the superconductor until it is either pinned by another pinning centre or leaves the superconducting domain. The pinning force therefore directly influences the critical current density j_c, since it is defined as the maximum current that a superconductor can carry without dissipation. So if the current flowing locally becomes larger than j_c, the excessive current will shift to regions where the electric current density j is still below j_c. Not only the electric energy determines whether a vortex will tear free of a pinning centre. The magnetic and thermal energies are equally important. Besides the pinning force, the pinning centre density influences the total electric critical current as well, resulting in the total critical current being proportional to both the pinning centre density ρ_p and the pinning force F_p:

$$I_c \propto F_p \cdot \rho_p. \tag{2.11}$$

2.5 Voltage-Current Dependence

In the dissipation free state a voltage drop is not observable due to the superconducting currents. That means that the total electric current I flowing is below the critical electric current I_c. On reaching the transition state at I_c, a measurable resistance occurs (see Fig. 2.6).

Mathematically, the observed behaviour is often conveniently modelled by a power-law dependence relating the voltage U to the total electric current I. The critical electric current I_c is determined using the arbitrary $1\,\mu\mathrm{V\,cm}^{-1}$ criterion: for a sample with voltage taps placed 5 cm apart, the critical current is reached when the total measured voltage reaches $5\,\mu\mathrm{V}$. This technique has not yet been standardised for *REBCO* coated conductors, only for Nb_3Sn, NbTi and BSCCO tapes [DKE07a, DKE07b, DKE06]. For the low-T_c superconductor material class, the criterion is actually $100\,\mathrm{nV\,cm}^{-1}$ due to their steeper transition. The power-law dependence with the power-law exponent n_p determining the steepness of the transition of the dependence of the scalar electric field E as a function of the I is:

$$E = E_c \left| \frac{I}{I_c} \right|^{n_p} \tag{2.12}$$

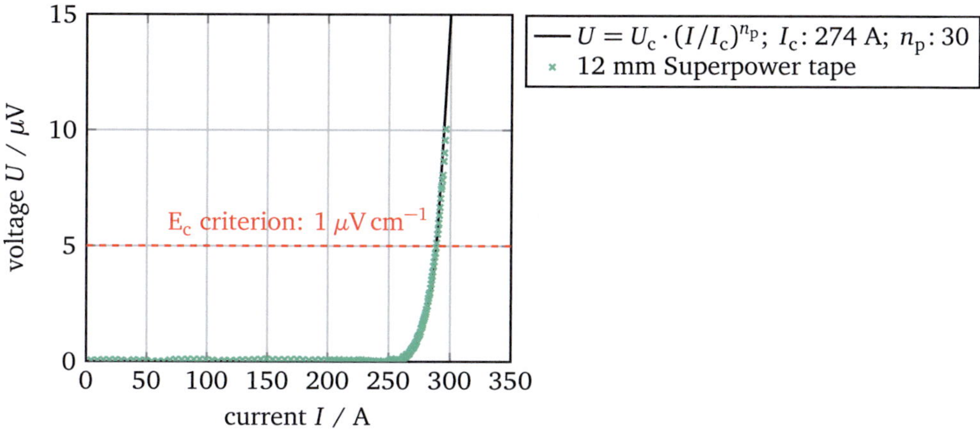

Figure 2.6: Superconducting U-I-characteristic measured on a 12 mm wide Superpower tape and fitted with a power-law dependency and the following parameters: $I_c=$ 274 A and $n_p=$ 30. The voltage taps were situated 5 cm apart, hence the $5\,\mu$V I_c criterion. The region where a notable resistance is measured, close to I_c, is called the transition state.

where the critical electric field E_c of around $10^{-4}\,\mathrm{V\,m^{-1}}$ is a typical value for high T_c-superconductors. The current-voltage characteristic can easily be fitted to measured data, see Fig. 2.6.

2.6 Loss Mechanisms in Superconductors

Despite the fact that ideal superconductors with applied electric transport currents or background magnetic fields insufficient to cause magnetic flux penetration show no dissipation, in measurements of Type-II superconductors, losses are often observed. Some of the losses stem from normal conducting components and others from super-conducting components. The different loss mechanisms will be explained subsequently. Their location of occurrence is roughly sketched in Fig. 2.7.

Hysteretic losses will be analysed in Cap. 2.6.1 and, they being an important source of losses, in great detail later on. Since superconducting wires or cables usually also comprise some normal conducting materials (see Cap. 2.7), a few more loss mechanisms are discussed, some of them solely prevailing to the normal conducting regions. The concept of flux creep (see Cap. 2.6.2) and flux flow (see Cap. 2.6.3) is introduced which is important for Type-II and, among those, for high-T_c superconduct-ors especially. The high-T_c superconductors are afflicted with flux creep to a greater degree since not only are their working temperatures usually higher but they also have shorter coherence lengths as compared to conventional superconductors that result in smaller pinning energies [BI93]. Coupling losses are important when a cable or wire consists of more than one superconducting filament (see Cap. 2.6.6). A similar loss mechanisms that is observed in superconductors with one filament as well is the eddy current loss (see Cap. 2.6.4). And lastly, ohmic losses in normal conducting regions are discussed (see Cap. 2.6.5).

2.6.1 Hysteretic Losses

When applying an electric transport current or, alternatively, a background magnetic field, magnetic flux will start penetrating the superconductor above a certain threshold, as discussed in Cap. 2.4. Supposing a certain magnetic flux is located in the volume of the superconductor enclosed in vortices, then reversing the electric transport current or the excitatory background magnetic field and applying an equally large but opposing current or field leads to the same magnetic flux penetrating the superconducting volume as initially, just with its polarity reversed. Since energy has to be expended in order to tear vortices from their pinning centres, a hysteresis of the magnetisation is observable and energy is dissipated [SM60].

In order to design superconducting applications, understanding hysteretic losses is vitally important since they are one of the most important loss factors resulting in heat dissipation in superconducting applications with transient electric currents or magnetic fields. This also makes them an important cost factor. Knowing where, spatially and temporally, energy is dissipated is also crucial for designing superconducting systems: finding hot spots and optimising geometries and materials in applications is necessary for maintaining operational stability on parameters like temperature, magnetic field

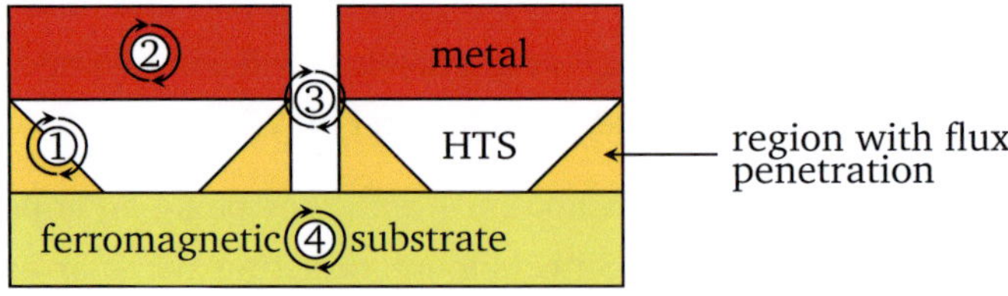

Figure 2.7: Schematic of a typical HTS coated conductor with locations of various loss mechanisms: 1) hysteretic losses in regions where magnetic flux has penetrated 2) eddy current losses in the normal conducting metal stabiliser 3) coupling losses between superconducting filaments (in multiple strand wires, cables or striated conductors) 4) ferromagnetic losses in the substrate.

and current density. Since parameters like magnetic field and electric current are interdependent, being able to simulate applications is necessary if prototyping via multiple trial and error iterations is to be avoided.

In isotropic and Type-I superconductors hysteretic losses are not observed as only Type-II superconductors show this loss mechanism. The losses are a direct result of the irreversibility of the flux pinning. They are equal to the integral of the area enclosed by the magnetisation-curve, with the energy dissipation per cycle Q a function of the magnetic field H and the sample magnetisation M:

$$Q = \int H \cdot dM = \int M \cdot dH. \tag{2.13}$$

Since the magnetisation only changes as a consequence of a change of magnetic field, the fact that the hysteretic losses are caused by temporal transients in the field is immediately obvious. If the magnetic field, either due to an ambient background field or due to the self field created by a transport current does not change, no hysteretic losses will be observable. Conversely, a change in magnetic field or an alternating electric current will lead to hysteretic losses.

2.6.2 Flux Creep

The initial vortex distribution forming due to the influences of either an electric transport current flowing or a background magnetic field being applied are influenced by the magnetic history of the material and the pinning force F_p as well as the pinning centre density ρ_p. If the superconducting state is stable and far from the critical parameters, that is, if the superconductor is not in the flux flow regime, the vortex distribution will not change significantly. This behaviour may also be thought of as the thermal energy not being sufficiently high enough to supply the activation energy to allow a flux vortex to tear free of its pinning centre. A vortex can still tunnel out of its position, however [FGLL93].

When energy of sufficient magnitude is supplied, either in thermal, electric or magnetic form, the statistical probability of the vortex overcoming the pinning force F_p and tearing free of its pinning centre is enlarged and eventually a vortex will be released from its initial position to move through the superconductor under the influence of the Lorentz force F_L until trapped again. Over time, the local magnetisation diminishes and a thermodynamically more stable state is achieved, with an electric current density j flowing that may be significantly lower than the critical current density j_c. After an infinite amount of time, the vortices will be distributed evenly in the material. The initial vortex distribution smears out over time. This thermally activated vortex movement is called flux creep.

Because flux creep is a thermally activated mechanism, it scales according to the Boltzmann law and the frequency with which the vortices will tear free is proportional to $\exp\left(-\Delta E_a/k_B T\right)$, with the Boltzmann constant k_B and the activation energy E_a, which depends on the electric current density j as the vortices break free at the critical current density j_c even at a temperature $T=0$ K: $E_a \propto \Delta(j_c - j)$.

Because of the temperature dependence, flux creep is particularly relevant for high-T_c superconductors. This is why, for high-T_c superconductors, the voltage-current characteristic is much smoother at the transition than for low-T_c superconductors.

2.6.3 Flux Flow

Flux flow is a state of constant flux movement, where the pinning force is either non-existent or too small to counteract the Lorentz force F_L. This means that the fluxons are in constant motion and thus steadily dissipate energy because transient magnetic fields induce electric fields. The electric field resulting from the vortices moving perpendicular to the transport current is anti-parallel to the electric field of the transport current. This constitutes a resistive voltage and power is dissipated [Tin96, pp. 162-176]. As is the case with flux creep, flux flow is more important in high-T_c superconductors, especially if the operating temperature is close to the critical temperature T_c. The higher the thermal energy of the system, meaning the closer the temperature is to the superconductor's critical temperature, the weaker the pinning forces will become. This is why even high-T_c superconductors perform better at lower temperatures [MU97, pp. 123 ff.].

2.6.4 Eddy Currents

When conductors (not only superconductors but also normal conductors) are exposed to varying magnetic fields, electric currents are induced. These eddies of current themselves cause magnetic fields. Since the induced electric currents and the resulting magnetic flux add to the magnetisation of the superconductor, eddy currents may lead to hysteretic losses (compare Cap. 2.6.1) [VLW04, pp. 71 ff.]. Because the eddy currents are superimposed on the possibly already established transport or shielding currents and may interact constructively or destructively, the local current and magnetic field distribution may be strongly influenced. This can lead to increased losses. Because the losses connected with eddy currents scale with the square of the frequency of the incident magnetic field, they can be identified in dissipation measurements.

Losses due to eddy currents may be reduced by special geometric designs. In cables, twisting of the strands with complete strand transposition reduces the effective area through which the flux passes, leading to reduced eddy currents. The twisting also leads to the eddy currents flowing in smaller loops, balancing out in the cross-over regions, further reducing the load on the conductor.

2.6.5 Ohmic Losses

Joule heating can occur with stabilised superconducting wires. If the outer layer is not superconducting but as in some coated conductors made of copper or other normal conducting materials, the current may not immediately transpose into the superconducting material. This leads to a voltage drop and consequently to ohmic losses due to resistive conduction. This effect most often occurs at contacts but can also be observed at defects in the superconductor. The influence of end effects is observable in $U(I)$-measurements as a linear component superimposed onto the power-law behaviour (compare with Cap. 2.5). Apart from these end effects, the normal conducting electrons inside the flux vortices generate ohmic losses according to Joule's first law with the energy dissipation per cycle Q, the total electric current I and the total electric resistance R:

$$Q = I^2 R. \tag{2.14}$$

2.6.6 Coupling Losses

The term coupling losses is usually not used to describe losses at engineered junctions in superconductors, for example when two cables are patched together. Instead, the term describes the losses occurring in the normal-conducting matrix between superconducting filaments. Often, all the filaments or strands of a cable or wire are transposed to reduce the eddy currents. If the strands are transposed but embedded in a conducting matrix, additional losses due to ohmic resistance will arise however, should the eddy currents flow through normal conducting parts. These losses are the so called coupling losses [Cam97].

2.7 Superconducting Materials

Superconductivity is not a rare phenomenon. In fact, almost half of the chemical elements are superconducting in pure form (see Fig. 2.8). Almost all of the elemental superconductors are Type-I superconductors whereas most compound superconductors are Type-II superconductors with only a few elemental Type-II superconductors like those mentioned in the beginning of Cap. 2. Prominent examples of Type-II superconductors are high-T_c superconductors of the oxocuprate class such as the rare earth barium oxocuprates, the barium strontium calcium oxocuprates and MgB_2, but also NbTi and Nb_3Sn. The latter two materials are low-T_c superconductors, and to date account for roughly 99 % of the total superconducting material market volume together with MgB_2 [Con].

REBCO is considered a high potential material because of the high critical temperature of around 90 K and the excellent in-field performance, meaning the material is able to conduct large amounts of electric current even when exposed to strong magnetic fields. Another advantage of *REBCO* as opposed to BSCCO is the absence of the silver matrix due to which the latter is very expensive. Even if no additional manufacturing costs are assumed the price for BSCCO will stay high. The high critical temperature allows liquid Nitrogen to be used as coolant for these materials instead of Helium which is considerably more cost-efficient. The raw-materials required in order to produce a superconducting *REBCO* tape are minuscule since the superconducting layer is only in the order of 1 μm thick. The *REBCO* tapes are mechanically more robust than the BSCCO variant due to their steel substrate which is important in high magnetic field applications where large forces may act on cables and tapes.

Superconducting materials are legion with superconductivity even being observed in organic compounds such as the Bechgaard salts [JMRB80] and alkali-doped fullerenes such as $RbCs_2C_{60}$ [BW96]. Niobium based superconductors are used in a lot of technical applications (see Cap. 2.7.1). MgB_2 is gaining importance and is introduced in Cap. 2.7.2 but does not reach as high critical currents as the oxocuprates while having a significantly lower critical temperature. Therefore, the research was focused on the promising material group of the oxocuprates, see Cap. 2.7.3.

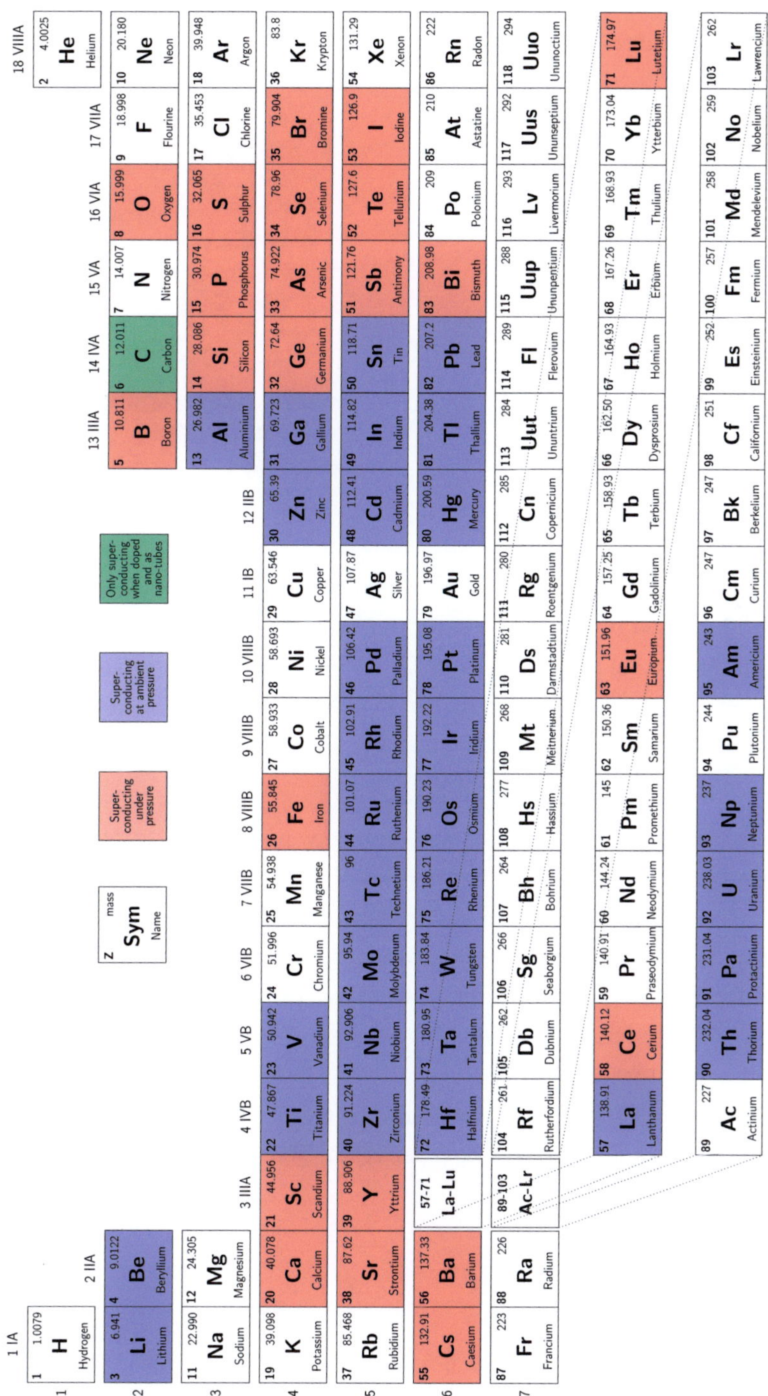

Figure 2.8: Periodic system of superconducting elements.

2.7.1 Niobium Based Superconductors

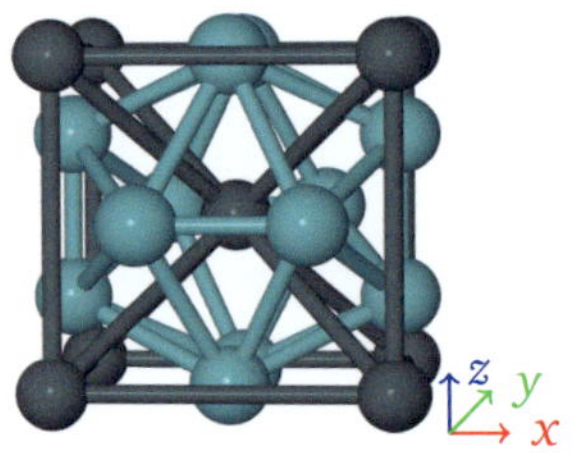

Figure 2.9: Crystal structure of Nb_3Sn, with the lattice parameter $a = 5.29$ Å.

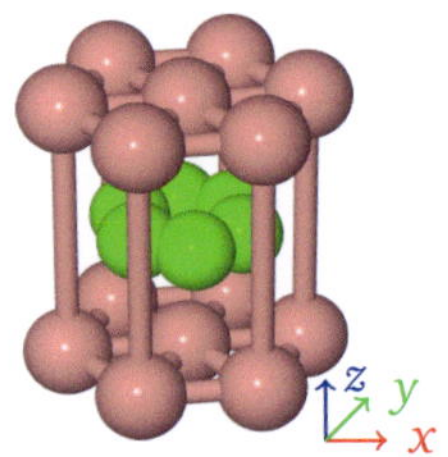

Figure 2.10: Hexagonal crystal structure of MgB_2, with lattice parameters $a = 3.086$ Å and $c = 3.5240$ Å.

Niobium even in elemental form being one of the Type-II superconductors (T_c of 9.2 K) is widely used in superconducting applications. Prominent among the Niobium based superconducting materials are NbTi and Nb_3Sn. NbTi is the most used material except in very specialised, high performance applications. It is also used due to its easy workability and hence cost-effectiveness [Bra09] and has a critical temperature of 9.8 K.

When even higher magnetic fields are required, Nb_3Sn is used, see Fig. 2.9. It is costly because of its complicated manufacturing process due to its extreme brittleness [JLL03]. It cannot be worked once synthesised and hence has to be shaped while in precursor form and then tempered; for coils, this manufacturing process is called wind and react [KTA$^+$96]. The crystal structure of Nb_3Sn is that of a body centred cubic of the β-Tungsten (also called A15) structure. Its upper critical field is 24 T and its critical temperature 23.2 K. Conductors made from both Nb_3Sn and NbTi consist of a multiplicity of thin superconducting strands, weaved together as opposed to coated conductors, whose superconducting layer is singly connected, see Cap. 2.7.3.

2.7.2 Magnesium Diboride

Although MgB_2 had been known for half a century and was even commercially available, its superconducting properties were only discovered in 2001 [NNM$^+$01]. That same year, MgB_2 has been used to manufacture superconducting wires [CFB$^+$01]. It has a moderately high critical temperature of $T_c = 39$ K. This means that liquid

Nitrogen cannot be used to cool MgB_2, one of the main advantages of *REBCO* (see Cap. 2.7.3).

Nevertheless, MgB_2 still has considerable potential due to the manufacturing process being much less complicated, faster and hence cheaper than in the case of *REBCO* or Nb_3Sn [VKS07]. MgB_2 cables are round as opposed to the high aspect ratio *REBCO* tapes which diminishes magnetisation losses. Also, liquid Helium is not necessarily required for cooling MgB_2 as conduction cooling may be used. Conductors made from MgB_2 consist, just like those made from Niobium based superconductors, of lots of thin strands combined into one wire. The wire need not only consist of MgB_2 strands. Different designs include copper for stabilisation or structural materials in order to improve the mechanical stability of the resulting wire. See Fig. 2.10 for the crystal structure of MgB_2.

2.7.3 Oxocuprates

Oxocuprates were the focus of intense research following their discover in 1986 [BM86]. Technically, two groups of oxocuprates are of particular interest: BSCCO and, above all, *REBCO*. BSCCO ($Ba_2Sr_2Ca_1Cu_2O_{8+\delta}$ with a T_c of 94 K and $Ba_2Sr_2Ca_2Cu_3O_{10+\delta}$ with a T_c of 107 K) is expensive due to the silver matrix enclosing the superconducting strands. Like MgB_2, BSCCO is usually manufactured using the powder in tube route.

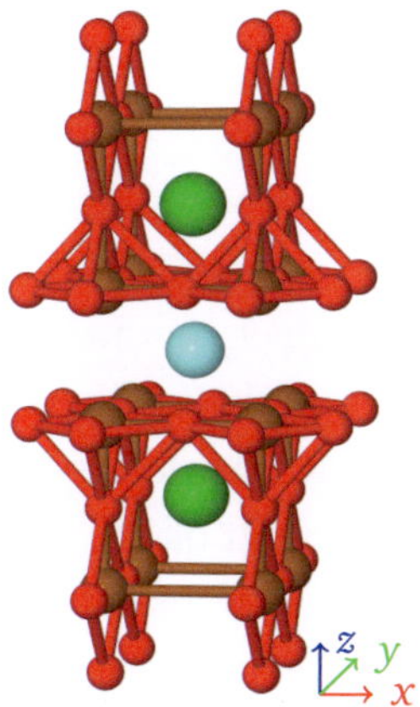

Figure 2.11: Perovskite based crystal structure of $YBa_2Cu_3O_{7-\delta}$, with lattice parameters $a = 3.8227$ Å, $b = 3.8872$ Å and $c = 11.6802$ Å.

Coated conductors, which sport superconducting layers made from *REBCO* are considered to have huge potential for technical applications. Not only because of their excellent physical properties such as high critical temperatures above the boiling point of liquid Nitrogen (T_c(YBa$_2$Cu$_3$O$_{7-\delta}$)$=93$ K) and high critical current density of about 10 kA mm^{-2} but also due to the minuscule amount of material needed and the resulting low material costs. For a schematic view of a coated conductor, refer to Fig. 2.12. The actual superconducting layer is only in the order of 1 μm thick, see Fig. 2.13.

Actual conductors are usually between 2 mm and 12 mm wide, leading to aspect ratios between 2000 and 12000. These high aspect ratios – also in combination with the single connectivity of coated conductors – are of importance when considering hysteretic losses in coated conductors. In perpendicular magnetic fields or when carrying electric transport currents, magnetic flux will start penetrating from the lateral edges where the magnetic field line density is highest. Due to the geometric layout and the high aspect ratio, the magnetic field in these regions is strongly intensified. This leads to much stronger magnetic flux penetration as for example the same transport current in a cylindrical conductor would. For more detailed information on material properties of coated conductors, refer to [Bar13, Cap. 2.3.3].

All important high current cable designs, both for fusion as well as for power applications, are based on coated conductors, as are most power application designs like transformers, fault current limiters and rotating machinery. This is why subsequently the coated conductors will solely be considered.

2.8 Ferromagnetic Materials

Ferromagnetic materials are of interest when investigating the properties of superconductors since the magnetic properties of both influence the magnetic field response of each other. Unlike superconductors which, in the Meißner state, are perfect diamagnets and thus expel magnetic flux from their volume, ferromagnets attract magnetic flux lines. This leads to the magnetic flux density being much higher inside ferromagnetic materials compared to vacuum or air. The ratio by which the magnetic flux density is amplified is governed by the relative magnetic permeability μ_r. For ferromagnetic materials, $\mu_r \gg 1$ whereas for superconductors $-1 <= \mu_r < 1$. Ideal superconductors show $\mu_r = -1$. Just like superconductors, ferromagnets have hysteretic behaviour: upon exposition to oscillating magnetic fields they dissipate energy as heat.

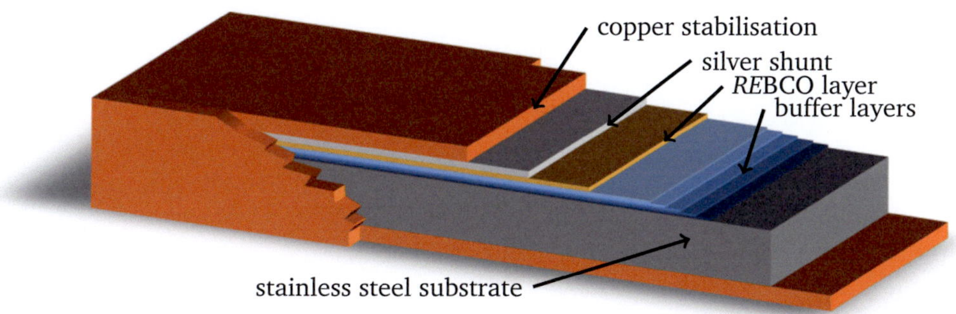

Figure 2.12: Schematic of the layered structure of a coated conductor. Instead of steel substrates, Rolling Assisted Biaxially Textured Substrates (RABiTS) are also used. These are mostly made from the ferromagnetic alloy $Ni_{95\,\%}W_{5\,\%}$. Those ferromagnetic substrates influence the hysteretic loss behaviour quite strongly, see Cap. 5.2. Note that the aspect ratio of the coated conductor shown is not representative. Reproduced with permission from [Bar13].

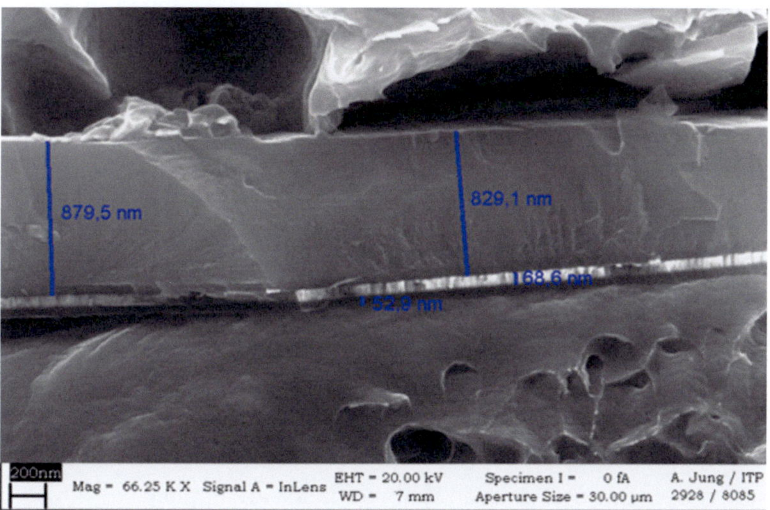

Figure 2.13: REM microscopy of a Superpower coated conductor tape. The superconducting layer is only about 850 nm thick while the buffer layers are even thinner, about 120 nm in total. For a schematic view of the layer structure, refer to Fig. 2.12. Picture courtesy of A. Jung, KIT.

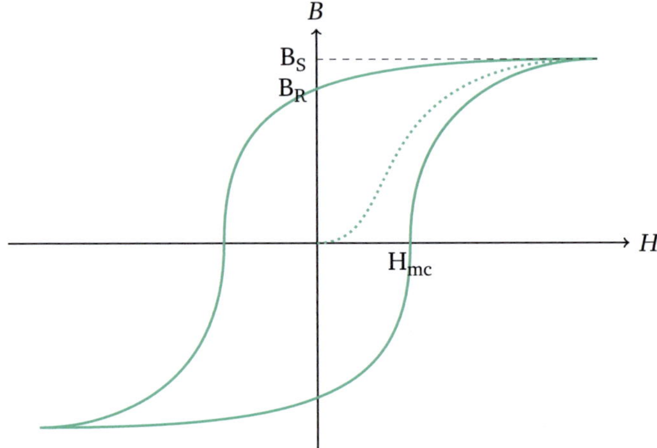

Figure 2.14: Hysteresis of ferromagnetic materials. The initial magnetisation is shown with a dotted curve. The hysteretic loss in ferromagnetic materials is proportional to the area enclosed by the $B(H)$ curve. A material with a higher remanence B_R, saturation flux density B_S and magnetic coercive field H_{mc} does not necessarily need to have a larger hysteretic loss as the area depends on both B_S and H_{mc} and if one is small, the resulting area integral is small as well.

The hysteretic behaviour of ferromagnetic materials is sketched in Fig. 2.14. The hysteretic losses in the ferromagnet depend on the area enclosed by the hysteresis curve. Generally speaking, the higher the saturation flux density B_S and the magnetic coercive field H_{mc}, the higher the losses. Higher saturation flux density B_S or magnetic coercive field H_{mc} do not automatically result in higher losses however: if one parameter remains low resulting in a narrow shape of the curve, the losses remain small. The remanence B_R denotes the amount of magnetic flux remaining in the sample after the outer magnetic field has been removed.

When the applied magnetic fields vary in amplitude, not only does the magnetisation of the ferromagnetic material change, but also the relative permeability. For all known materials, if the applied magnetic field approaches infinity, the relative permeability tends to unity. This accounts for the fact that the material is saturated and the magnetisation does not change even if higher fields are applied.

2.8.1 Nickel

Nickel is one of three elements besides iron and cobalt showing ferromagnetic properties at room temperature. At cryogenic temperatures, Gadolinium ($T_c = 289$ K),

Terbium ($T_c = 220$ K), Dysprosium ($T_c = 85$ K), Holmium ($T_c = 20$ K) and Erbium ($T_c = 32$ K) become ferromagnetic as well. Being an elemental ferromagnetic material eases deposition considerably as Nickel can simply be electroplated from a wet solution. Most of the other ferromagnetic materials are compounds and thus plating methods would have to be developed. Additionally, the corrosion resistance of Nickel is very good and it is a well understood material. All of the above predestine Nickel as a material for initial investigations of the interactions of superconductors and ferromagnetic materials.

In order to be able to account for Nickel in the numerical simulations, its permeability and loss function have to be modelled correctly. With data from [GvV$^+$07], a numerically efficient formula was constructed for the dependence of the relative magnetic permeability μ_r on the local magnetic field amplitude H:

$$\mu_r(H) = -75.5 \cdot \{\arctan(0.00017 \cdot H)\}^{0.9924} + 119.19 \tag{2.15}$$

The loss function is defined as the sum of the integrals over the area of the ferromagnetic domains:

$$Q_{FM} = \sum_{i=1}^{n} \int_{S_{FM}} f_{FM} \, dS_{FM_i} \tag{2.16}$$

where n is the number of ferromagnetic domains present. Then we have the surface S_{FM} of the ferromagnetic domain in question. The arbitrary loss function f_{FM} is defined as:

$$f_{FM} = \begin{cases} Q_{sat} \cdot \left(\frac{\vec{B}}{\vec{B}_{sat}}\right)^2 & \vec{B} \leq \vec{B}_{sat} \\ Q_{sat} & \vec{B} > \vec{B}_{sat} \end{cases} \tag{2.17}$$

with the local flux density amplitude $B = \mu_0 \cdot \mu_r \cdot H$, the saturation flux density $B_{sat} = 500$ mT and the saturation energy dissipation per cycle $Q_{sat} = 2.75$ MJ m^{-3} per cycle.

We used electroplating to deposit Nickel on the samples and the solution used was a Watts bath. For concentration information, see Tab. 2.2.

Reaching the desired thickness of the Nickel coating by electroplating is quite simple using Faraday's law of electrolysis:

$$m_d = \frac{I \cdot t \cdot M}{z \cdot F}. \tag{2.18}$$

Concentration of reagents required for electroplating Nickel	
Reagent	Concentration / $\mathrm{g\,l^{-1}}$
Nickel Sulfate $NiSO_4$	$240 - 310$
Nickel Chloride $NiCl_2$	$20 - 50$
Nickel Dihydrate $Ni(HCOO)_2 \cdot 2H_2O$	≈ 45
Boric Acid H_3BO_3	$20 - 40$
Cobalt(II) sulfate $CoSO_4$	≈ 4.5
Formaldehyde CH_2O	≈ 2.5
Ammonium Sulphate $(NH_4)_2SO_4$	≈ 0.75

Table 2.2: Concentration of reagents required for electroplating Nickel using a Watts bath. Other bath compositions are also possible (Nickel Sulfamate, all-Chloride, Sulfate-Chloride, all Sulfate, hard Nickel). The electromagnetic properties of the deposited Nickel have to be checked when other electroplating methods are used.

Here, m_d is the mass to be deposited, calculated from the volume and density ρ of Nickel:

$$m_\mathrm{d} = w \cdot l \cdot d_\mathrm{sub} \cdot \rho. \tag{2.19}$$

where w is the width of the shielded area, l is the length of the sample and d_sub is the shield thickness. Care has to be taken to compute the correct volume: usually two sides of a conductor will have to be coated. The time t designates the period for which a current I has to flow and the valence of Nickel z (for all practical considerations: 2). Finally, F is the Faraday constant, which is $96\,485\ \mathrm{C\,mol^{-1}}$.

In general, it is advisable to chose a current that is rather too low than too high, since a lower current produces more uniform results. Additionally, a lower voltage is needed in order to maintain the chosen current. Through the coating gaining in thickness and thus in resistivity and the ions in solution depleting, the total conductivity of the Watts bath is reduced with time and the voltage needs to be increased in order to balance this effect. Since the voltage of any given current source is limited, keeping the deposition current low ensures a stable process without reaching the voltage limit.

Additionally, it is considered to be beneficial to arrange the sample in a way that the wide face of the coated conductor faces the electrode. This allows uniform deposition due to the symmetrical arrangement. If one side were situated closer to the electrode, Nickel would primarily be deposited there.

2.8.2 Nickel-Wolfram Alloys

There are two distinctive approaches to manufacturing coated conductors at present: using electropolished non-magnetic stainless steel substrates with a number of buffer layers or using rolling assisted biaxially textured substrates (RABiTS), which were first proposed in 1996 [GNB$^+$96, GNK$^+$97]. The material class used for RABiT substrates is a NiW-alloy. Various stoichiometries such as $Ni_{9\%}W$ and others have been used [DGL$^+$05, SZL$^+$07], but the most common variant is $Ni_{5\%}W$ which is ferromagnetic [DTG$^+$03, DGL$^+$05, MGC$^+$06, GBF$^+$06, MYV$^+$07, MFR$^+$08].

In order to achieve the required crystallisation structure of *REBCO* with low-angle grain boundaries, the superconducting layer has to be deposited on a suitably prepared surface. The RABiTS process provides such a surface in a very efficient manner since rolling is a fast mechanical process whereas depositing thin film buffer layers physically or chemically is slow in comparison [GNB$^+$96, GNK$^+$97]. Using NiW-alloy based substrates reduces the number of required buffer layers considerably and provides a favourable structure while speeding up the production process. Because of the interaction of ferromagnetic material and superconductors, a RABiTS based coated conductor behaves differently from a conductor with non-magnetic substrate (see Cap. 5.2).

In order to model the physical behaviour of the $Ni_{5\%}W$-alloy, fitting functions reproducing the $\mu_r(B)$ and the $Q(B)$ response were employed which in turn are based on measured data from [MYU$^+$08].

2.8.3 Further Ferromagnetic Materials

Nickel as a ferromagnetic material does not have a particularly high relative magnetic permeability nor a particularly low loss function. Materials with high permeability and low hysteretic losses would be ideally suited for ferromagnetic shielding (see Cap. 2.8.4) since the flux rerouting would be stronger and the addition to the overall losses smaller. In Tab. 2.3, an overview of interesting materials is provided. It should be noted that further investigations recording detailed ferromagnetic behaviour have to be conducted as single values for initial relative permeability, saturation flux density and hysteresis loss are not sufficient for modelling and understanding the complex interactions between superconductors and ferromagnetic materials. Also, measurements at cryogenic temperatures are required in order to know how these materials behave in liquid Nitrogen or Helium.

Typical properties of several soft magnetic materials at room temperature		
Material	Initial Relative Permeability μ_r	Saturation Flux Density B_S / T	Hysteresis Loss per Cycle / $\mathrm{J\,m^{-3}}$
Commercial iron ingot	150	2.14	270
Silicon-iron (oriented)	1400	2.01	40
45 Permalloy	2500	1.60	120
*Ferroxcube A	1400	0.33	40
*Ferroxcube B	650	0.36	35
Sendust	35000	1.0	18
Hiperco	>3000	2.4	200
Permendur	700	2.45	300
Cold-rolled Si-Steel	1500	2.0	44
78.5 Permalloy	10000	1.07	20
3.8-78.5 Cr-Permalloy	12000	0.80	20
3.8-78.5 Mo-Permalloy	20000	0.85	20
*Supermalloy	100000	0.8	20
Hipernik	4500	1.6	10
*TDK PE 22	1800	0.51	3.16
*TDK PE 90	2200	0.53	2.4

Table 2.3: Properties of ferromagnetic materials at room temperature adapted from [CR10, p. 822] and [Hos52, pp. 204, 214, 231, 236] and from a material data sheet on PE 22 & PE 90 (v.D07EA2 2010.05.17) by TDK Corporation, Tokyo, Japan. A comprehensive study at cryogenic temperatures would be most interesting. Asterisks mark possibly interesting materials.

2.8.4 Ferromagnetic Shielding

While superconductivity is usually suppressed by high magnetic fields and magnetic impurities in the material, in 1999, it was predicted that ferromagnetic materials could also have a beneficial effect on superconductors by increasing the critical current if placed in certain geometric configurations [GUF99]. Properly shaped and positioned ferromagnetic shields are supposed to redirect flux and thus permit a metastable state to assert itself. By delaying the flux entry they should permit overcritical currents flowing in the superconductor volume up to a certain threshold [GSF00, Gen02, JJF02, JBH05, GRKN09]. Later, the concept was extended to reduce hysteretic losses in the superconductor [GVPv09]. However, since the ferromagnetic material also dissipates energy in oscillating magnetic fields, a trade-off has to be found between reducing losses in the superconductor and increasing losses in the ferromagnetic material. Also, great care has to be taken regarding placement and geometry of the ferromagnet lest only detrimental effects occur.

3 Modelling Superconductors

Even though only direct measurements provide final confirmation regarding the outcome of experiments, employing modelling in a scientific setting is often beneficial if practicality and convenience are considered: modelling complex systems may not only speed up investigations considerably, it may furnish tools to investigate otherwise inaccessible observables. The latter in particular is extremely helpful in discerning the influences of convoluted variables, helping to understand basic principles.

While analytic models provide exact solutions for a given set of boundary conditions (see Cap. 3.1), numerical methods are able to consider almost arbitrary problems (see Cap. 3.2).

3.1 Analytic Models

The earliest analytic model able to account for the behaviour of Type-II superconductors is the critical state model discussed in Cap. 3.1.1 [Bea62, BD62, Bea64, Lon63]. Simplifying, it assumes constant critical current which to a degree is remedied by incorporating vector-field-dependent critical current densities (see Cap. 3.1.2). Geometries like the technically highly interesting coated conductors tapes and ellipse-shaped conductors are taken into account by the Norris and Brandt models in Cap. 3.1.3.

3.1.1 Critical State Model

The basic assumption of the macroscopic critical state theory, also called Bean theory, accounts for the fact that hysteresis was observed in superconducting samples [Bea62, BD62, Bea64, Lon63]. Opposed to the microscopic theory focusing on vortices and pinning centres however, the critical state model is a field theory. This means that the material is treated as isotropic and instead of the Lorentz force F_L and the pinning centre density ρ_p, a macroscopic quantity has to be found to account for the observed behaviour. This quantity is the critical electric current I_c. It relates to the magnetic field as:

$$\mu_0 \cdot j_c = |\nabla \vec{B}|. \tag{3.1}$$

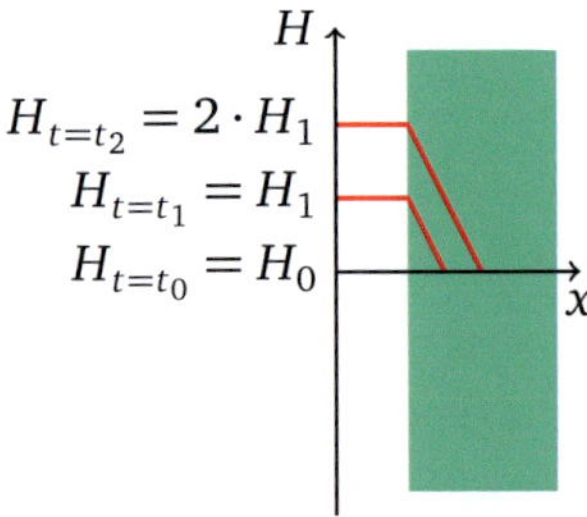
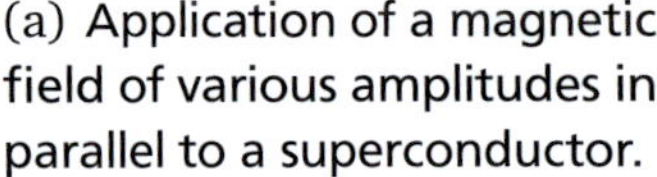

(a) Application of a magnetic field of various amplitudes in parallel to a superconductor.

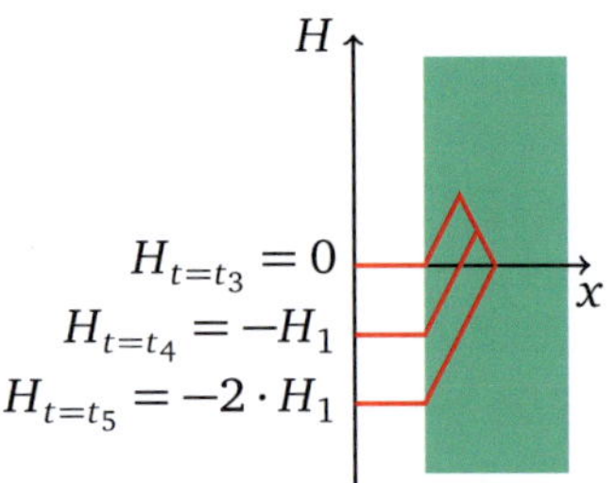

(b) Subsequent application of a magnetic field of inverted direction.

Figure 3.1: A slab of superconductor never having experienced magnetic fields or currents is subjected to a magnetic field of various amplitudes and directions: at $t = t_0$ there is no applied background magnetic field and null field inside the superconductor. At $t = t_1$ a field H_1 is applied, leading to a linearly diminishing field inside the superconductor with $dH/dx = j_c$. After having applied a field of $2 \cdot H_1$ at $t = t_2$, the total measurable magnetisation will only become zero at $t = t_4$ with an applied field of $-H_1$.

In that respect, the critical state theory and the macroscopic theory are alike: once the local critical current density j_c is reached, the current penetrates further into the superconducting volume and the magnetic field changes accordingly.

Areas where no magnetic field is present are initially devoid of current in the virgin superconducting state, meaning they have not experienced exposure to magnetic fields or transport currents since entering the superconducting state. The hysteresis stems from the superconductor remembering the history of magnetisation: suppose a superconductor was exposed to a magnetic field of sufficient magnitude to penetrate its volume but insufficient to cause total penetration. The shielding currents induced by said field will continue to flow due to the dissipation free state even if the excitation field is removed.

The superconductor will retain its magnetisation and will act as a permanent magnet as long as neither the superconducting state collapses nor the polarity of the exciting field is reversed. Should the latter be the case and a field of half the magnitude but opposing direction to the primary excitation field be applied and subsequently removed, the summary magnetisation of the superconductor will be zero. This intriguing behaviour is due to the fact that during the application of the second field, the flux trapped in the superconductor volume has to change polarity. This is achieved by vortices moving out of and vortices of opposing polarity moving into the material. Even though the total magnetisation of the superconductor will be null, the current profile of the sample will not be the same as in the virgin state: inside the superconductor, two shielding currents of equal magnitude will flow contra-directionally (see Fig. 3.1).

3.1.2 Vector-field-dependent Critical Current Densities

While Bean acknowledged the fact that the critical current density j_c should be dependent on the local magnetic field in his original paper [Bea62], he still assumed it to be constant in his theory. While in reality j_c should be a vector due to the anisotropy of many superconductors, it is in fact assumed to be scalar in almost all analytical and numerical models. The first model incorporating a field dependence was the Kim model [KHS62], relating j_c to the scalar magnetic flux density B as

$$j_c(B) = \alpha/B. \tag{3.2}$$

Apart from this very simple approach, depending on the material and the geometry under observation, it is sometimes necessary to take into account the components of the magnetic field independently. Especially when dealing with quasi two-dimensional material structures as in oxocuprates [ITTT87]. A reasonable agreement with measurements sufficient for most cases is achieved by using the following equation with the intrinsic zero-field critical current density j_{c0}, the parameter of anisotropy k, the parallel flux density component $B_\parallel$, the perpendicular flux density component $B_\perp$, the characteristic flux density B_0 and the field-dependence exponent b [GK06]:

$$j_c(x,y) = \frac{j_{c0}}{\left(1 + \frac{\sqrt{k^2 \cdot B_\parallel{}^2 + B_\perp{}^2}}{B_0}\right)^b}. \tag{3.3}$$

The most thorough descriptions of the critical currents in anisotropic superconductors with isotropic pinning sites being manufactured today requires taking care of reproducing the predominant direction of pinning centres. This is possible using linear combinations with multiple components of elliptical dependencies [BGL92, DBT$^+$04, GLG$^+$07]. Most prevalent coated conductors may be described using the following set of equations very successfully [PVGv11], where the incident magnetic field angle θ of magnetic flux density $\vec{B}$ is taken into account:

$$j_c(|\vec{B}|(x,y),\theta) = \{j_{c,ab}[|\vec{B}|f_{ab}(\theta)]^m + j_{c,c}[|\vec{B}|f_c(\theta)]^m + j_{c,i}[|\vec{B}|f_i(\theta)]^m\}^{\frac{1}{m}} \tag{3.4}$$

with

$$j_{c,ab}[|\vec{B}|f_{ab}(\theta)] = \frac{j_{0p}}{[1 + |\vec{B}|f_{ab}(\theta)/B_{0ab}]^b}$$

$$j_{c,c}[|\vec{B}|f_c(\theta)] = \frac{j_{0p}}{[1 + |\vec{B}|f_c(\theta)/B_{0c}]^b}$$

$$j_{c,i}[|\vec{B}|f_i(\theta)] = \frac{j_{0i}}{[1 + |\vec{B}|f_i(\theta)/B_{0i}]^a}$$

(3.5)

and

$$f_{ab}(\theta) = \begin{cases} f_{ab0}(\theta) & \forall \theta \in [-90° + \delta_{ab}, 90° + \delta_{ab}] \\ f_{ab\pi}(\theta) & \forall \theta \notin [-90° + \delta_{ab}, 90° + \delta_{ab}] \end{cases}$$

(3.6)

and finally

$$f_{ab0}(\theta) = \sqrt{\cos^2(\theta - \delta_{ab}) + u_{ab}^2 \sin^2(\theta - \delta_{ab})}$$

$$f_{ab\pi}(\theta) = \sqrt{v^2 \cos^2(\theta - \delta_{ab}) + u_{ab}^2 \sin^2(\theta - \delta_{ab})}$$

$$f_c(\theta) = \sqrt{u_c^2 \cos^2(\theta - \delta_c) + \sin^2(\theta - \delta_c)}$$

$$f_i(\theta) = \sqrt{\cos^2(\theta - \delta_i) + u_i^2 \sin^2(\theta - \delta_i)}.$$

(3.7)

All the parameters $(j_{0p}, j_{0i}, B_{0ab}, B_{0c}, B_{0i}, a, b, \delta_{ab}, \delta_c, \delta_i, u_{ab}, u_c, u_i, v)$ with the exception of the exponent $m = 8$ are material specific and the functions have to be fitted to measured data. For exemplary data, see [PVGv11].

Closely related to the simulation of critical current densities being dependent on the local magnetic field amplitude are critical current densities depending on mechanical stress fields. The twisted stacked tape cable for example, see Cap. 7, with its longitudinal screw-like geometry is affected by compressive and tensile stress. If the twist pitch becomes too small, reversible and – at even smaller twist pitches – irreversible degradation of the critical current density is observed. This behaviour has been investigated in [TMBB10]. The formulas required to model the dependence of the critical current on the mechanical stress were implemented but the differences in results compared to not taking into account said effects are negligible for all practical considerations. At the same time, the computational effort required rises unreasonably due to the increased complexity. Because the dependence of the critical current density is material specific and negligible below a threshold (in this case: less than $100 \cdot 10^{-3}$ %), there is little benefit in taking into account the effects. Especially so since the dependence has to be measured anyway and simply keeping the twist pitch above the threshold of permanent material damage removes the requirement for simulation altogether. This is generally true for power applications, a mechanical stress analysis might be in order for magnet applications.

3.1.3 Analytic Model for Coated Conductors

Both the Norris [Nor70] as well as the Brandt [Bl93] model is able to predict the magnetic field and the electric current distribution around and inside superconductors for a number of geometries. Only the latter is able to also account for background applied magnetic fields however. Most important for technical applications are the solutions for the strip and the ellipse geometry as those resemble *REBCO* and BSCCO tapes, respectively.

For applied transport current, the hysteretic power loss P in a strip and an ellipse is [Nor70]:

$$P = \frac{f \cdot I_c^2 \cdot \mu_0}{\pi} \begin{cases} \{(1 - I/I_c) \cdot \ln(1 - I/I_c) + (1 + I/I_c) \cdot \ln(1 + I/I_c) - (I/I_c)^2\} & \text{(Strip)} \\ \{(1 - I/I_c) \cdot \ln(1 - I/I_c) + (2 - I/I_c) \cdot (I/I_c)\} & \text{(Ellipse)} \end{cases} \tag{3.8}$$

For an applied background field the losses in a strip are:

$$P = 4f \cdot \mu_0 \cdot a^2 \cdot J_{S,c} \cdot H_a(2H_c/H_a) \cdot \ln(\cosh(H_a/H_c)) - \tanh(H_a/H_c), \tag{3.9}$$

with the frequency f, the half width of the superconducting strip a, the applied magnetic background field H_a and the critical magnetic field H_c, in the Brandt model defined as $H_c = J_{S,c}/\pi$ where the critical sheet current $J_{S,c}$ is defined as $J_{S,c} = I_c/(2a)$.

It is good practise to compare numerical and experimental results obtained from geometries resembling strips or ellipses to the analytical Norris and Brandt solutions for initial comparison since striking deviations usually indicate erroneous data. It is fortunate that both the technically interesting coated conductors and also the first generation conductors made from BSCCO are described quite well by these geometric abstractions.

In the following chapters, when considering coated conductor tapes, electric current as well as magnetic field profiles that are results of numerical simulations are often compared with analytic formulas. The analytical solutions for the sheet current J_S as well as the magnetic field amplitude H are again taken from Ref. [BI93]. The analytic formula for the case of applied electric transport current relies on the magnetic field penetration depth position γ, which is defined in this case as:

$$\gamma = a \cdot \sqrt{1 - \left(\frac{I_a}{I_c}\right)^2} \tag{3.10}$$

and is:

$$J_S(x) = \begin{cases} (2 \cdot J_{S,c}/\pi) \cdot \arctan\sqrt{\frac{a^2-\gamma^2}{\gamma^2-x^2}} & |x| < \gamma \\ J_{S,c} & \gamma < |x| < a. \end{cases} \tag{3.11}$$

For the case of applied background magnetic field, γ is defined as:

$$\gamma = a/\cosh(H_a/H_c) \tag{3.12}$$

with H_c defined as in Cap. 3.1.3 and a further temporary constant c is introduced for readability:

$$c = \tanh\left(\frac{H_a}{H_c}\right). \tag{3.13}$$

Finally, the formula for the sheet current is:

$$J_S(x) = \begin{cases} (2 \cdot J_{S,c}/\pi) \cdot \arctan\left(\frac{c \cdot x}{\sqrt{\gamma^2-x^2}}\right) & |x| < \gamma \\ J_{S,c}\frac{x}{|x|} & \gamma < |x| < a. \end{cases} \tag{3.14}$$

The combined load case of applied electric transport current and applied magnetic background field is basically a superposition of the two cases and is calculated as:

$$J_S(x) = J_{S,c}[j(x+w, a+w, \gamma) + p_p \cdot j(-x-w, a-w, \gamma)] \tag{3.15}$$

with c defined as in Eq. 3.13, a similarly defined constant ζ for the fraction of the applied critical electric current I_c:

$$\zeta = \frac{I_a}{I_c}, \tag{3.16}$$

the constants b_1:

$$b_1 = a \cdot \sqrt{1-\zeta^2} \cdot \sqrt{1-c^2} + \zeta \cdot c, \qquad (3.17)$$

and b_2:

$$b_2 = a \cdot \sqrt{1-\zeta^2} \cdot \sqrt{1-c^2} - \zeta \cdot c, \qquad (3.18)$$

and a field penetration depth position γ resulting therefrom:

$$\gamma = \frac{b_1 + b_2}{2}, \qquad (3.19)$$

and further constants w:

$$w = \frac{b_1 - b_2}{2}, \qquad (3.20)$$

p:

$$p = \sqrt{|x^2 - \gamma^2| \cdot (a - \gamma)}, \qquad (3.21)$$

and the function $j(x, a, \gamma)$:

$$j(x, a, \gamma) = \begin{cases} 1 & \gamma \leq x \leq a \\ (1/\pi) \cdot \mathrm{arccot}\left(\frac{\gamma^2 - a \cdot x}{p}\right) & x \leq |\gamma| \\ 0 & -\infty < x \leq -\gamma \end{cases} \qquad (3.22)$$

The formula for the magnetic field profile in the case of applied transport current relies on γ as defined before in Eq. 3.10 and is:

$$H(x) = \begin{cases} 0 & |x| < \gamma \\ \frac{H_c \cdot x}{|x|} \cdot \mathrm{arctanh}\sqrt{\frac{x^2 - \gamma^2}{a^2 - \gamma^2}} & \gamma < |x| < a \\ \frac{H_c \cdot x}{|x|} \cdot \mathrm{arctanh}\sqrt{\frac{a^2 - \gamma^2}{x^2 - \gamma^2}} & |x| > a. \end{cases} \qquad (3.23)$$

The magnetic field profile in the case of applied background magnetic field, with γ defined as in Eq. 3.12, is:

$$H(x) = \begin{cases} 0 & |x| < \gamma \\ H_{\mathrm{c}} \cdot \operatorname{arctanh} \frac{\sqrt{x^2 - \gamma^2}}{c \cdot |x|} & \gamma < |x| < a \\ H_{\mathrm{c}} \cdot \operatorname{arctanh} \frac{c \cdot |x|}{\sqrt{x^2 - \gamma^2}} & |x| > a. \end{cases} \tag{3.24}$$

And the formula in the case of both applied electric transport current and background magnetic field, with γ defined as in Eq. 3.19, is:

$$H(x) = H_{\mathrm{c}}[h(x + w, a + w, \gamma) - p_{\mathrm{p}} \cdot h(-x - w, a - w, \gamma)], \tag{3.25}$$

with

$$h(x, a, \gamma) = \begin{cases} 0 & |x| \leq \gamma \\ \frac{x}{2 \cdot |x|} \cdot \operatorname{arctanh}\left(\frac{p}{a \cdot x - \gamma^2}\right) & \gamma \leq |x| \leq \infty. \end{cases} \tag{3.26}$$

The polarity constant p_{p} in Eq. 3.15 and in Eq. 3.25 may take the values ± 1 and governs whether the electric current dominated or the magnetic field dominated case is calculated. The following cases determine which value p_{p} takes:

$$p_{\mathrm{p}} = \begin{cases} -1 & \text{if } I < I^*(H_{\mathrm{a}}) \\ +1 & \text{if } I > I^*(H_{\mathrm{a}}) \end{cases} \tag{3.27}$$

which is equivalent to

$$p_{\mathrm{p}} = \begin{cases} -1 & \text{if } H_{\mathrm{a}} < H_{\mathrm{a}}{}^*(I_{\mathrm{a}}) \\ +1 & \text{if } H_{\mathrm{a}} < H_{\mathrm{a}}{}^*(I_{\mathrm{a}}). \end{cases} \tag{3.28}$$

The star denotes the electric current I generated by the applied magnetic field H_{a} and vice versa.

3.1.4 Analytic Model for Bilayered Coated Conductors

Just as is the case with single coated conductors, simple bilayer structures consisting of a superconducting and a ferromagnetic layer may also be described using analytical formulas [Maw08]. Assuming the critical state, infinitesimally thin layer structures and infinite permeability of the ferromagnetic layer allows the solution of this particular problem using conformal mapping.

The field lines resulting from an applied magnetic field are calculated as follows (see Ref. [Maw08]):

$$H(x) = \begin{cases} 0 & |x| < \gamma \\ \Re(\mathfrak{H})(x + i\epsilon) & \gamma \leq |x| \leq a \end{cases} \tag{3.29}$$

with γ defined as in Eq. 3.12, the positive infinitesimal ϵ tending towards zero and the complex field $\mathfrak{H}$:

$$\Re(\mathfrak{H}) = 2 \cdot H_c/\pi \cdot \arctanh \left(\sqrt{\frac{\sqrt{a^2 - \gamma^2} \cdot (\eta + a)}{a \cdot (\eta + \sqrt{a^2 - \gamma^2})}} \right) - \frac{\sqrt{a \cdot \sqrt{a^2 - \gamma^2} \cdot (\eta + a) \cdot (\eta + \sqrt{a^2 - \gamma^2})}}{(a + \sqrt{a^2 - \gamma^2}) \cdot \eta} \tag{3.30}$$

with the transformed variable η, used to calculate the conformal mapping:

$$\eta = i \cdot \sqrt{(x + iy)^2 - a^2}. \tag{3.31}$$

As mentioned before, the analytic solutions serve as reliable test cases. They behave like a numerical simulation with $n_p = \infty$ so for high n_p values, say $n_p = 1000$, the results should be almost indistinguishable. In the case of a ferromagnetic substrate being present, the same is true for the permeability: since in the analytic formulation, the permeability is assumed infinite, the higher the permeability in the numerical simulations, the more the results should be in accord.

3.2 Numerical Finite Element Models

Experimental work suffers from two major drawbacks: physical effects may seldom be observed isolatedly and preparing samples and conducting experiments may be expensive due to costly materials. Analytic formulas on the other hand provide exact and elegant solutions to problems but are often unsuited for arbitrary problems such as complex geometries. Numerical solutions however are able to answer problems of almost unlimited complexity, at least within the constraints of computational power at disposal. Variables are observable that may not be directly accessible experimentally and numerical simulations are in most cases much faster than experiments. Using parametrisation and additional degrees of freedom, it is possible to optimise a given problem for a set of parameters; a task most cumbersome with experimental work. Geometric optimisation could for example be used to find the best possible layout without prototyping.

The Finite-Element-Analysis is a numerical method for solving partial differential equations. The simulation domain Ω is discretised into an arbitrary amount of elements. These elements have finite size and the subsequent description of the problem using a finite amount of degrees of freedom lent the method its name. The size of the elements and, consequently, the amount of degrees of freedom determine the speed with which the problem will be solved. For large geometries, extensive parameter sets or excessively fine detail, the computation time may become extremely high, ranging from hours to days or even weeks. It is therefore imperative to optimise the simulation. It is always possible to trade accuracy for speed but with intelligent design, it is also possible to reduce complexity without sacrificing accuracy.

A number of numerical formulations and models exist for simulating superconductors, including the T-Ω-formulation based on the current vector potential T and the scalar magnetic potential Ω [HSM$^+$91, AMBM98], the A- or A-V-formulation based on the magnetic vector potential [NTFA89, Seb94, Pri97, BMDH99, CFD99, CFD99, HYT$^+$00, ST10] and the magnetic vector potential critical state model [Cam07, Cam09]. The finite element model employed in almost all of the simulations in this work is based on the well established H-formulation [BV83, KTK$^+$01, KHY$^+$03, PMC$^+$03, HCC06, BGM07, NAW09]. During the investigations detailed subsequently, the extensions to the H-model allowing for non-linear $\vec{B}$-$\vec{H}$-relations as in [NAW09] were further developed. A considerable speed-up and improvement of stability was achieved (see Cap. 3.2.5). Additionally, in order to be able to account for end-effects such as contact resistances, a new boundary condition was developed, see Cap. 3.2.6.

For finding the parameters to model the superconducting tapes used during the experimental work correctly, a model based on the A-V-formulation was used. The parameters that were extracted from the measurements are critical current, n_p-value and the modified Kim-model [GK06] parameters field-dependence exponent b and parameter of anisotropy k (compare Cap. 2.5 & 3.1.2). The model based on the A-V-formulation was used since it solves steady-state problems much faster than the H-formulation, which is sufficient for the purpose of the investigation. Since the model was only used once a thorough explanation is skipped. It shall be mentioned that imposing boundary conditions is not as straightforward as in the H-model [HC10]. The same applies for calculating hysteretic losses [ST10]. In all numerical investigations, the commercially available FEM software COMSOL was used [COM13].

3.2.1 Governing Equation

The H-formulation uses the magnetic field as its state variable and is based on the following set of equations:

$$\nabla \times \vec{H} - \frac{\partial \vec{D}}{\partial t} = \vec{J},$$ (3.32)

with the electric current $\vec{J}$, the variable of time t and, since we are only investigating problems in the low frequency regime such as power applications and the displacement current $\vec{D}$ being assumed negligible and hence zero, the equation is therefore simplified to:

$$\nabla \times \vec{H} = \vec{J}.$$ (3.33)

Further, the Maxwell–Faraday equation:

$$\nabla \times \vec{E} = \frac{\partial \vec{B}}{\partial t},$$ (3.34)

and Gauß's law for magnetism is used:

$$\nabla \cdot \vec{B} = 0.$$ (3.35)

The electric field $\vec{E}$ is related to $\vec{J}$ with the resistivity ρ as:

$$\vec{E} = \rho \vec{J},$$ (3.36)

and $\vec{B}$ to H:

$$\vec{B} = \mu_0 \mu_{\rm r} \vec{H}.$$ (3.37)

Substituting Eq. 3.36 and 3.37 into the Maxwell-Faraday Eq. 3.34:

$$\nabla \times (\rho \nabla \times \vec{H}) = -\frac{\partial (\mu_0 \mu_{\rm r} \vec{H})}{\partial t}$$ (3.38)

and Eq. 3.37 into Gauß's law for magnetism, Eq. 3.35, we obtain:

$$\nabla \cdot (\mu_0 \mu_r \vec{H}) = 0.$$ (3.39)

Since the equation system containing both Eq. 3.38 and 3.39 is over-constrained, the approach by Kajikawa et al. is used and the divergence of Eq. 3.38 is taken, yielding [KHY$^+$03]:

$$\nabla \cdot (\nabla \times (\rho \nabla \times \vec{H})) = \nabla \cdot \left(-\frac{\partial (\mu_0 \mu_r \vec{H})}{\partial t} \right).$$ (3.40)

The divergence of a curl being equal to zero and exchanging:

$$\nabla \cdot \left(-\frac{\partial (\mu_0 \mu_r \vec{H})}{\partial t} \right) = \left(-\frac{\partial (\nabla \cdot (\mu_0 \mu_r \vec{H}))}{\partial t} \right) = \left(-\frac{\partial (\nabla \cdot (\vec{B}))}{\partial t} \right),$$ (3.41)

it is immediately obvious that if $\nabla \cdot (\vec{B}) = 0$ at the timestep zero t_0, then $\nabla \cdot (\vec{B}) = 0 \ \forall \ t > t_0$. Strictly speaking, this constraint is only true at the timestep t_0 and numerical errors may accumulate over time. This leads to divergence of the solution. Appropriate measures such as tweaking of the temporal and spatial discretisation and of the numerical solver (the Newton method was generally used) can mediate this behaviour. It follows that one constraint of the equation system has to state that:

$$\nabla \cdot (\vec{B}(t_0)) = 0.$$ (3.42)

Eq. 3.40 is the governing equation being solved for the state variable H by the numerical simulation software. Since we use COMSOL, our formula has to be slightly rearranged to conform with COMSOL notation, which usually denotes the state variable u. For two-dimensional simulations:

$$u = \begin{bmatrix} H_x \\ H_y \end{bmatrix}$$ (3.43)

The equation being solved is:

$$e_a \frac{\partial^2 u}{\partial t^2} + d_a \frac{\partial u}{\partial t} + \nabla \cdot \Gamma = f_s.$$ (3.44)

The mass coefficient e_a is always zero in the (first order) H-model and the damping coefficient d_a is defined as:

$$d_\mathrm{a} = \begin{bmatrix} \mu_0\mu_\mathrm{r} & 0 \\ 0 & \mu_0\mu_\mathrm{r} \end{bmatrix}.$$
(3.45)

The source term f_s is also zero while the conservative flux vector Γ is defined as:

$$\Gamma = \begin{bmatrix} 0 & E_z \\ -E_z & 0 \end{bmatrix}.$$
(3.46)

3.2.2 Applying Background Magnetic Field and Transport Current

The constitutive equations having been outlined in Cap. 3.2.1, some boundary conditions are still needed in order to apply excitatory influences. Dirichlet boundary conditions imposed on the circumferential simulation domain boundary $\partial\Omega$ of the simulation environment are suited to the application of magnetic field, with some complications also for transport currents or a combination of both. The boundary condition specifies the values the solution may take in that particular position.

Even when not seeking to apply excitatory loads to the simulation on the outermost boundary, setting boundary conditions is required in order to minimise influences of the finite simulation environment on simulation results. In order to compensate for the finite size of the simulation geometry, the asymptotic boundary value has to be set to be consistent with any other boundary conditions. But still, the simulation environment should not be too small in order to allow the numerical error to dissipate in the air (or vacuum) domain surrounding the superconductor. If no magnetic field and no electric transport current is applied, the Dirichlet boundary condition is trivially zero. If however the magnetic field generated by possibly flowing electric transport currents that were also imposed as boundary conditions need to be compensated, we need to account for the influence of the electric transport currents on the magnetic field and define the magnetic field perturbation H_p on the simulation domain boundary $\partial\Omega$.

Assuming $\partial\Omega$ is sufficiently far from the conductor (at least ten times the width of the widest superconductor dimension) and only far field effects need to be allowed for, the incident magnetic field on the boundary may be treated as if generated by a point source:

$$H_{\text{p}} = I_{\text{ext}}(t)/\left(2\pi \cdot \sqrt{x^2 + y^2}\right) \tag{3.47}$$

where the applied transport current I_{ext} is the total applied transport current in the longitudinal (in this instance, in the cartesian coordinate z) direction and $r = \sqrt{x^2 + y^2}$ (with the cartesian coordinate x and the cartesian coordinate y) is the distance from the conductor (also see Fig. 3.2). The externally applied magnetic field H_{ext} is the applied magnetic field to be imposed on the boundary.

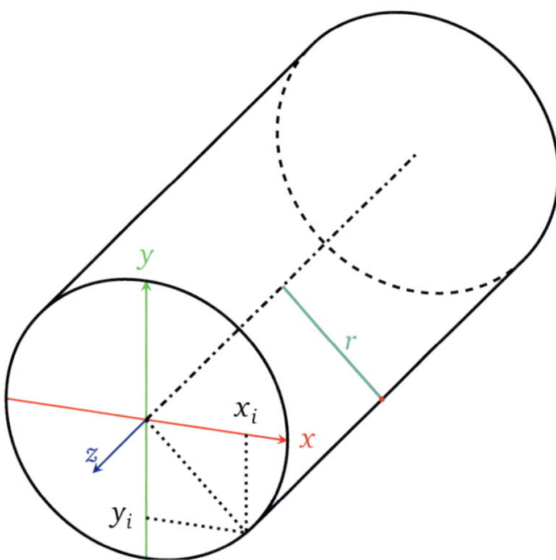

Figure 3.2: If the boundary is sufficiently far away from the superconductor (roughly ten times the width of the widest superconductor dimension), the latter may be treated as having a dimensionality of zero in the cross-section which enables calculation of the field given solely the distance from the origin where the conductor is situated and the total electric current flowing through the conductor. For any point P_{i} on $\partial\Omega$, the distance is simply $r = \sqrt{x_{\text{i}}^2 + y_{\text{i}}^2}$.

Since the magnetic field $\vec{H}$ is used as the state variable, this means that $\vec{H}$ has to be constrained component-wise to specific values as a function of time on the boundary. With the superconducting cable, tape or wire arbitrarily oriented with the wide face parallel to the xz-plane, the direction of perpendicular field with respect to the conductor is along the y-direction. Since for most investigations, the case of perpendicular field is the most interesting, the following example shows a background magnetic field being applied along the y-axis-direction. Naturally, any other orientation is possible as well. The component-wise definition of the Dirichlet boundary condition is:

$$H_x = -H_{\mathrm{p}} \cdot y / \sqrt{x^2 + y^2}, \tag{3.48}$$

$$H_y = -H_{\mathrm{ext}}(t) + H_{\mathrm{p}} \cdot x / \sqrt{x^2 + y^2}. \tag{3.49}$$

If a three-dimensional geometry is considered, there is usually no reason to apply magnetic field along the longitudinal direction of the tapes or wires, so the third component is zero:

$$H_z = 0. \tag{3.50}$$

When investigating power-applications, sinusoidal current and field profiles are usually used so the following definitions would be typical for the applied background magnetic field:

$$H_{\mathrm{ext}} = \mathrm{rm}(t) \cdot \mu_0 \cdot H_{\mathrm{a}} \cdot \sin\left(2\pi f t\right) = \mathrm{rm}(t) \cdot \mu_0 \cdot H_{\mathrm{a}} \cdot \sin\left(\omega t\right) \tag{3.51}$$

with the pulsatance ω defined as $\omega = 2\pi f$. Since the simulation converges much faster if a consistent set of initiation variables has been set (remember that $\nabla \cdot (\vec{B}(t_0)) = 0$, compare Eq. 3.42), the internal ramping function $\mathrm{rm}(f)$ of the FEM software is used, resulting in an applied current profile as in Fig. 3.3. It is important that not only $H_{\mathrm{ext}}(t_0) = 0$ but also $dH_{\mathrm{ext}}(t_0)/dt = 0$ so an adequate ramp function has to be used. The COMSOL ramping function is set to have a starting value of zero and two continuous derivatives. This speeds up the initial convergence considerably.

Given the case of transport current, possibly flowing in several conductors, integral boundary conditions provide an easier possibility to apply the loads as the field distribution does not have to be known in advance. The coupled case where the applied current distributes self-consistently according to the self-field effect is achieved by defining one integral boundary condition encompassing all the conductor surfaces:

$$I = \sum_{i=1}^{n} \int j \cdot dS_n. \tag{3.52}$$

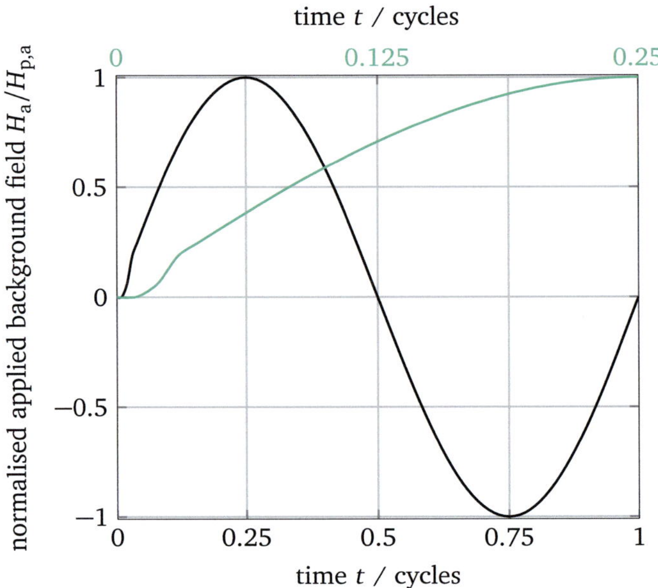

Figure 3.3: Typical magnetic background field profile applied in investigations of power applications. Shown is the applied background field normalised by the peak applied field. Transport current boundary conditions are similar but apply an electrical current instead of a magnetic field, of course. Using a ramping function to not only ensure $H_\text{ext}(t_0) = 0$ but also $dH_\text{ext}(t_0)/dt = 0$ leads to a significant speed-up as the initial numerical convergence is much faster; the green plot shows the first quarter of the whole profile enlarged.

Imposing a different electric current in each conductor is achieved by using one integral boundary condition per conductor (uncoupled case) where the current I_n flowing in a particular singly connected superconducting domain with the S_n is:

$$I_n = \int j \cdot dS_n.$$ (3.53)

The total applied current I_ext is then equal to the sum of the current in all superconducting domains. The applied transport current function is very similar to the function used to apply background magnetic fields in power applications:

$$I_\text{ext} = \text{rm}(t) \cdot I_\text{a} \cdot \sin(\omega t)$$ (3.54)

with the applied electric transport current I_a.

3.2.3 Material Laws

Apart from the equations mentioned in Cap. 3.2.1, initial conditions satisfying Eq. 3.39 and a set of boundary conditions and/or constraints, the model requires material laws. For superconducting domains, the $\vec{E}$-$\vec{J}$-relationship has to be defined. Most modelling applications assume isotropic tensors as does the model used. Some recent studies [BL01, Cam11, CWDC11] found that may not be sufficient and imply that an anisotropic tensor might be required. For the time being however, the following dependence for ρ is used:

$$\rho = \frac{E_c}{j_c} \cdot \left| \frac{j}{j_c} \right|^n \tag{3.55}$$

with the critical electric field E_c. While in Eq. 3.55, the critical current density j_c is described as a scalar, in general it depends on the local magnetic field (see Cap. 3.1.2) and the position, due to variations in the manufacturing uniformity.

3.2.4 Calculation of Hysteretic Loss

Once the magnetic and electric field profiles as well as the current distribution are calculated, extracting the hysteretic losses merely requires post-processing of the data. Since the air domain also present in the numerical simulations does not carry any current, the current distribution is physically meaningful inside of the superconducting domains only. Due to numerical reasons, an infinitesimally small current is present in the air region. The field distribution however is also interesting outside the superconducting domains since it is required in order to study background field effects on one particular superconducting domain.

The method used to extract the hysteretic losses in one or all superconducting domains depends slightly on the applied signal. For non-periodical loads the whole signal is taken into consideration. But when investigating steady-state dissipation and observing a periodic signal in a superconductor, the first half-cycle is disregarded, because it shows initial magnetisation effects and is not representative.

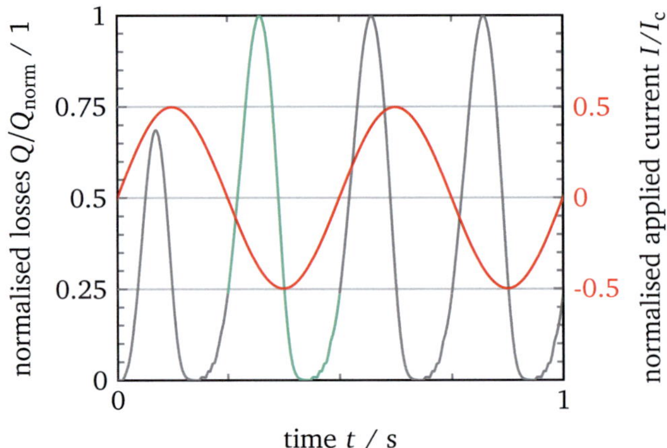

Figure 3.4: Exemplary normalised loss profile of a coated conductor with an electric transport current load of 50 % I_c. The applied electric transport current is plotted in red. The first half-cycle shows less dissipation because of the magnetisation from virgin state. For steady-state investigations, only the second half-cycle marked in green is therefore considered. Subsequent cycles provide no new information. Note the phase difference between peak applied current and maximum dissipation.

In the case of a sinusoidal signal for example, the integral over the second half is taken. So with the frequency f, the energy dissipation per cycle Q in J/m^3/cycle for all n superconducting domains is:

$$Q = 2 \cdot \sum_{i=1}^{n} \int_{1/2f}^{1/f} dt \int_{S} \vec{J} \cdot \vec{E} \, dS_i. \tag{3.56}$$

For other kinds of time-dependent signals such as ramps, the whole time range has to be considered:

$$Q = \sum_{i=1}^{n} \int dt \int_{S} \vec{J} \cdot \vec{E} \, dS_i. \tag{3.57}$$

In subsequent numerical simulations, usually only the axial (along the tape) component of the electric current and electric field are taken into account as the others are negligible.

3.2.5 Modelling Ferromagnetic Material

When seeking to simulate ferromagnetic materials using the H-model, it has to be taken into account that not only does the permeability have an influence on the magnetic field distribution, but the magnetic field strength at a given position also determines the local permeability of the ferromagnetic material. On top of that, since the ferromagnetic material also shows hysteretic behaviour, the additional losses in the ferromagnetic domains have to be taken into consideration.

The dependence of the permeability on the local magnetic field amplitude is incorporated with a material law while the inclusion of the ferromagnetic material's dynamic influence on the magnetic field distribution is more intricate. In order to model ferromagnetic materials, consider the Maxwell-Faraday Eq. 3.38 rewritten as in [NAW$^+$10]:

$$
\mu_0 \underbrace{\frac{\partial(\mu_{\mathrm{r}}(|\vec{H}|)\vec{H})}{\partial t}}_{\text{compare Eq. 3.59}} + \underbrace{\rho\nabla\times\vec{J}}_{\text{compare Eq. 3.63}} = 0 \tag{3.58}
$$

If the first summand of Eq. 3.58 is expressed as:

$$
\mu_0\frac{\partial(\mu_{\mathrm{r}}(|\vec{H}|)\vec{H})}{\partial t} = \mu_0\underbrace{\frac{\partial\mu_{\mathrm{r}}(|\vec{H}|)}{\partial t}\vec{H}}_{\text{compare Eq. 3.60}} + \mu_0\mu_{\mathrm{r}}(|\vec{H}|)\frac{\partial\vec{H}}{\partial t}, \tag{3.59}
$$

where again the first summand is conveniently rearranged for 2D geometries by neglecting $\mu_0\vec{H}$ for the time being as:

$$
\begin{aligned}
\frac{\partial\mu_{\mathrm{r}}(|\vec{H}|)}{\partial t} &= \frac{\partial\mu_{\mathrm{r}}(|\vec{H}|)}{\partial\vec{H}}\cdot\frac{\partial\vec{H}}{\partial t}\\[4pt]
&\xRightarrow{\vec{H}=\sqrt{H_x^2+H_y^2}}\frac{\partial\mu_{\mathrm{r}}(|\vec{H}|)}{\partial\vec{H}}\left(\frac{\partial\vec{H}}{\partial H_x}\cdot\frac{\partial H_x}{\partial t} + \frac{\partial\vec{H}}{\partial H_y}\cdot\frac{\partial H_y}{\partial t}\right)\\[4pt]
&= \frac{\partial\mu_{\mathrm{r}}(|\vec{H}|)}{\partial\vec{H}}\left(\frac{1}{2}\cdot\frac{2H_x}{\sqrt{H_x^2+H_y^2}}\frac{\partial H_x}{\partial t} + \frac{1}{2}\cdot\frac{2H_y}{\sqrt{H_x^2+H_y^2}}\frac{\partial H_y}{\partial t}\right)\\[4pt]
&= \frac{\partial\mu_{\mathrm{r}}(|\vec{H}|)}{\partial\vec{H}}\left(\frac{H_x}{\vec{H}}\frac{\partial H_x}{\partial t} + \frac{H_y}{\vec{H}}\frac{\partial H_y}{\partial t}\right)\\[4pt]
&= \underbrace{\frac{\partial\mu_{\mathrm{r}}(|\vec{H}|)}{\vec{H}\partial\vec{H}}}_{\equiv f}\left(H_x\frac{\partial H_x}{\partial t} + H_y\frac{\partial H_y}{\partial t}\right).
\end{aligned} \tag{3.60}
$$

Reinserting Eq. 3.60 into Eq. 3.59 yields:

$$\mu_0 f \vec{H} \left(H_x \frac{\partial H_x}{\partial t} + H_y \frac{\partial H_y}{\partial t} \right) + \mu_0 \mu_{\mathrm{r}}(|\vec{H}|) \frac{\partial \vec{H}}{\partial t} \tag{3.61}$$

Separating Eq. 3.61 into single component form:

$$
\begin{aligned}
x: \quad & \mu_0 f H_x \left(H_x \frac{\partial H_x}{\partial t} + H_y \frac{\partial H_y}{\partial t} \right) + \mu_0 \mu_{\mathrm{r}}(|\vec{H}|) \frac{\partial H_x}{\partial t} \\
& = \mu_0 \left(f H_x^2 + \mu_{\mathrm{r}}(|\vec{H}|) \right) \frac{\partial H_x}{\partial t} + \mu_0 f H_x H_y \frac{\partial H_y}{\partial t} \\
y: \quad & \mu_0 f H_y \left(H_x \frac{\partial H_x}{\partial t} + H_y \frac{\partial H_y}{\partial t} \right) + \mu_0 \mu_{\mathrm{r}}(|\vec{H}|) \frac{\partial H_y}{\partial t} \\
& = \mu_0 \left(f H_y^2 + \mu_{\mathrm{r}}(|\vec{H}|) \right) \frac{\partial H_y}{\partial t} + \mu_0 f H_x H_y \frac{\partial H_x}{\partial t}
\end{aligned}
\tag{3.62}
$$

The second summand of Eq. 3.58 is better written as

$$\rho \nabla \times \vec{J} = \rho \begin{pmatrix} \frac{\partial \vec{J}_z}{\partial y} - \frac{\partial \vec{J}_y}{\partial z} \\ \frac{\partial \vec{J}_x}{\partial z} - \frac{\partial \vec{J}_z}{\partial x} \\ \frac{\partial \vec{J}_y}{\partial x} - \frac{\partial \vec{J}_x}{\partial y} \end{pmatrix} = \rho \begin{pmatrix} \frac{\partial \vec{J}_z}{\partial y} \\ -\frac{\partial \vec{J}_z}{\partial x} \\ 0 \end{pmatrix} \tag{3.63}$$

which if reinserted in Eq. 3.58 together with Eq. 3.62 results in the final set of equations:

$$
\begin{aligned}
\mu_0(\mu_{\mathrm{r}}(|\vec{H}|) + H_x^2 f) \frac{\partial H_x}{\partial t} + \mu_0 H_x H_y f \frac{\partial H_y}{\partial t} + \rho \frac{\vec{J}_z}{\partial y} &= 0 \\
\mu_0(\mu_{\mathrm{r}}(|\vec{H}|) + H_y^2 f) \frac{\partial H_y}{\partial t} + \mu_0 H_x H_y f \frac{\partial H_x}{\partial t} + \rho \frac{\vec{J}_z}{\partial x} &= 0.
\end{aligned}
\tag{3.64}
$$

Remembering Eq. 3.43, we have to slightly alter our partial differential equation set-up since the dynamic ferromagnetic influence needs to be included in either f_{s} or in d_{a}. Eq. 3.64 can be inserted directly into d_{a}:

$$d_{\mathrm{a}} = \begin{bmatrix} (\mu_{\mathrm{r}} + \frac{\partial \mu_{\mathrm{r}}}{\partial H} H_x^2 / |\vec{H}|) \cdot \mu_0 & \mu_0 \cdot H_x H_y \cdot \frac{\partial \mu_{\mathrm{r}}}{\partial H} / |\vec{H}| \\ \mu_0 \cdot H_x H_y \cdot \frac{\partial \mu_{\mathrm{r}}}{\partial H} / |\vec{H}| & (\mu_{\mathrm{r}} + \frac{\partial \mu_{\mathrm{r}}}{\partial H} H_x^2 / |\vec{H}|) \cdot \mu_0 \end{bmatrix} \tag{3.65}$$

but the differential $\partial\mu_{\mathrm{r}}/\partial|\vec{H}|$ has to be calculated beforehand, since the FEM software is incapable of providing the solution on-the-fly.

In the course of the investigations, it was discovered that the stability of the above solution leaves a lot to be desired and the simulations often do not converge in a timely manner, sometimes not at all. The simulations are quite often unstable and tend to diverge without actually triggering an error, therefore producing erratic results or running indefinitely.

Instead of using the damping coefficient to account for the dynamic ferromagnetic influence, it is also possible to rewrite the equations so that the ferromagnetic influence is accounted for in the source term. This approach has the additional advantage of not having to evaluate the computationally expensive derivation $\partial\mu_{\mathrm{r}}/\partial|\vec{H}|$ and the FEM software provides direct access to $\partial\mu_{\mathrm{r}}/\partial t$. So by simply forcing:

$$f_{\mathrm{s}} = \begin{bmatrix} -\frac{\partial\mu_{\mathrm{r}}}{\partial t} H_x \\[2mm] -\frac{\partial\mu_{\mathrm{r}}}{\partial t} H_y \end{bmatrix} \tag{3.66}$$

not only is the convergence and reliability of the numerical simulations increased considerably, the simulation is also sped up by about 15 % to 50 %. When using the source term to simulate the ferromagnetic material, the damping coefficient has to be reset to the initial configuration:

$$d_{\mathrm{a}} = \begin{bmatrix} \mu_0\mu_{\mathrm{r}} & 0 \\ 0 & \mu_0\mu_{\mathrm{r}} \end{bmatrix}. \tag{3.67}$$

Lastly, the analytic function itself used to model the ferromagnetic material's response to the magnetic field also influences the speed of the simulation. For a qualitative comparison of the fitting, see Fig. 3.5. The exponential fitting function could be expanded to better fit the measured data but even a simple exponential fitting function of the form $a\cdot\exp(x/b)^c + d$ is computationally about one order of magnitude slower than a comparable trigonometric function. While using the optimised fitting function does not result in the speed up of 1000 % unfortunately, the total simulation time does decrease by about 10 %, which is considerable.

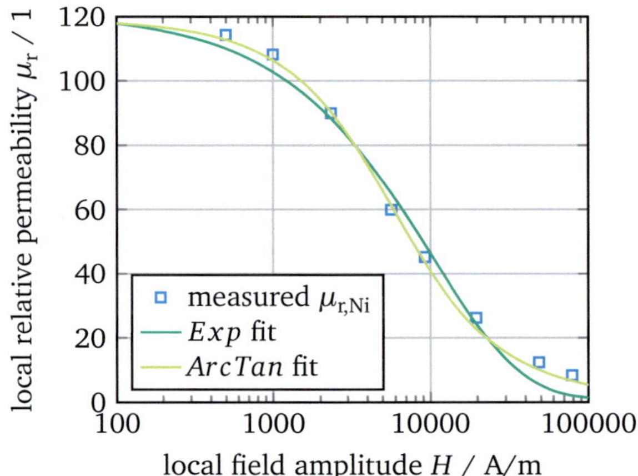

Figure 3.5: Comparison of different analytic functions to describe the $\mu_r(H)$ dependency. The exponential fitting function $a\exp(x/b)^d + e$ and the trigonometric fitting function $a(\arctan(bx)^c) + d$ are compared. Note the considerably better fit using the arcus tangens function. On top of the fitting function being a qualitatively better fit, it is also evaluated faster by one order of magnitude. This results in the FEM simulation running about 10 % faster.

3.2.6 Contact Resistances

The model outlined subsequently is a three-dimensional expansion of the model presented earlier. It is capable of not only simulating electric transport currents in three dimensions but also background magnetic fields. Additionally, without loss of generality nor disruption of translational symmetry, it is possible to take into account end-effects such as contact resistances.

Contact resistances are important in short samples as their influence may completely mask or suppress self-field effects. Even though they may be relatively small in long length samples, the contact resistances determine current distribution and hence electric behaviour in short length cables. Hence, it is highly interesting to be able to discern between the various influences. Also, in order to understand measurements better, modelling the contact resistance end-effects helps to deconvolute the two effects.

In order for the simulation of the end-effects to not influence the primary simulation environment and thus disrupt translational symmetry (when for example only one twist pitch is simulated), contacts cannot be included as distinct geometric objects in the primary simulation environment. Since the physical behaviour of the contacts themselves is not under scrutiny, a howsoever constructed analytic consideration would be preferable, ideally in the form of a zero-dimensional boundary condition, imposed as a constraint.

For a cable consisting of multiple superconducting wires or tapes, two suitable boundary conditions come to mind: instead of floating the electric potential on an arbitrary but equal level for all contacts, finding a gauge and adding an arbitrary offset to the electric potential for each superconducting tape representing the contact resistance of that particular tape. The electric potential offset of a contact resistance ΔU_i is calculated from the product of the (measured) contact resistance R_i and the current flowing through that particular conductor I_i:

$$U_i = R_i \cdot I_i. \tag{3.68}$$

If the common electric potential is fixed and U_i can be calculated from the sum of the resistance of the contact R_i and due to the self field effect R_c, then the electric current I_i flowing in a particular conductor i is:

$$I_i = \frac{U_i}{(R_i + R_c)}. \tag{3.69}$$

For the DC case, the following approach is valid: the electric potential of the contact is extracted from the boundary surface perpendicular to the translational symmetry by averaging over the electric field component normal to the surface. In the model developed here the translational symmetry is always along the z-component. The averaging is achieved by utilising an averaging operator of the FEM software:

$$U_i = \langle E_{z,i} \rangle \cdot L_i \tag{3.70}$$

and multiplying with the length of the simulated conductor L_i.

Another possibility to construct a suitable boundary condition is by calculating the equivalent resistivity and adding a constant value to each conductor according to its contact resistance. Calculation of the resistivity of the conductor is possible if firstly the size of each discretised element along the cross-section is known and secondly the resistivity of each element is known. Given these criteria, the conductor may be treated as a series of parallel connections of resistances and thus the total resistivity may be calculated. However in reality due to the material specific peculiar magnetic field and electric current distributions inside a superconductor most of the elements situated in the centre of the conductor will have zero resistance at low fields. Since the power-law is employed as a material law in the numerical simulations, this need not necessarily be a problem per se, as the power-law will always have a finite value, however small. Unluckily, for infinitesimally small values, convergence of the solution tends to become exceedingly slow and the simulation takes very long to complete, if it is able to converge steadily at all. If a finite minimum resistivity is ensured the simulation is much more stable. This is achieved by using a function to select the bigger of two input variables with the first variable being the positive infinitesimal ϵ and the second being the resistivity ρ. For two-dimensional problems and in the DC case, this leads to:

$$U_i = \frac{\int j_z dS_i}{\int \frac{1}{\max(\rho,\epsilon)} dS_i}. \tag{3.71}$$

One such boundary condition has to be constructed for each conductor i present.

Unfortunately, while both boundary conditions work theoretically, each suffers from different limitations. The parallel resistivity integration method's results rely heavily on the chosen value of ϵ. If one were to run the electric field averaging method first and use the result to gauge the parallel resistivity integration method, in subsequent runs, a correct solution could be obtained much faster. However, this approach is impractical. The electric field averaging method suffers the same affliction as the original parallel resistivity integration method without the complication of the *max* function: its convergence is exceedingly slow and hence as good as unusable in power applications with oscillating fields where long relaxation times of the simulation cannot be tolerated. Relaxation in this context means that an excitation load is ramped up quickly and then held constant (this is the so called relaxation) so that the numerical errors may dissipate. The longer (in amount of time-steps) the simulation is left sitting with constant loads, the better the result will resemble the physical behaviour.

If the already mentioned solution of a zero-dimensional boundary cannot be created, a remedy is the creation of a second simulation environment disconnected from the first (the domains remain entirely separated). For an example demonstrating such a configuration, please refer to Fig. 7.13. The only aspect which is of interest in this environment is the total resistance of arbitrarily sized domains representing the contacts. The magnetic field surrounding these contacts and the contact geometry should be of no consequence. A separate air domain for each contact would be even

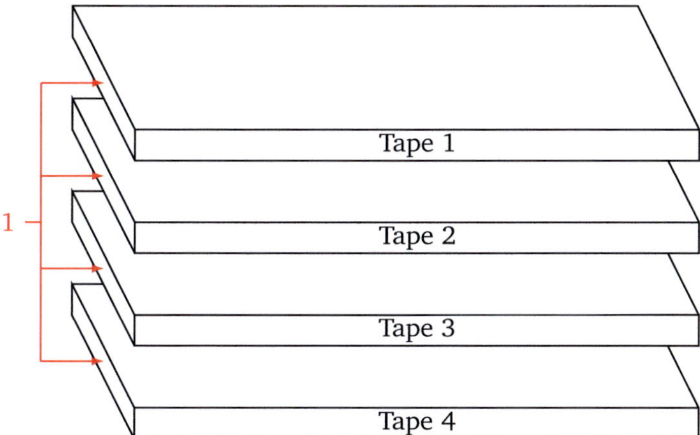

Figure 3.6: If no contact resistance is simulated, the current is applied at point 1 and since all coated conductor tapes are on the same potential the current will distribute according to the self-field effect only.

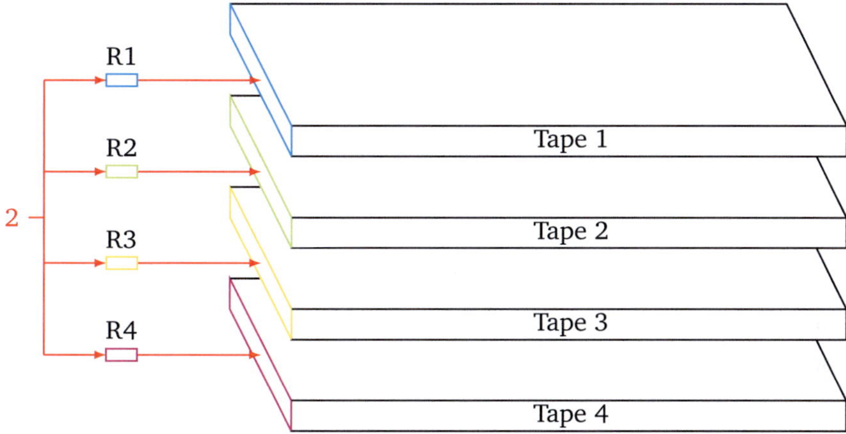

Figure 3.7: If a contact is simulated, the current is applied at point 2 and the contact resistances are added in series to the resistances of the tapes. Since now the potential is only the same before the resistors, both the voltage drop due to the contact resistance as well as that due to the self-field effect is taken into account.

more ideal. Then, the only complication lies in ensuring the continuity of current flows. This is achieved by integrating operators on each tape and contact surface perpendicular to the translational symmetry and global constraints forcing the currents to be equal in a contact and its corresponding tape via two corresponding integrating operators. The only thing left to do then is to assign the correct resistance to each contact domain in order to model the contacts correctly. The set-up is outlined in Fig. 3.7.

One additional restriction to keep in mind regards the skin effect as with increasing frequency and decreasing conductivity the current tends to flow on the outside (the skin) of any conductor. For quasi-static loads this is of little importance but when considering high frequency applications the dimensions of the contact domains may become important. It has been mentioned above that the domain size of the contact resistances may be chosen arbitrarily as long as the resistivity of the domain is calculated accordingly in order to result in the required total contact resistance. If the domain is large, the resistivity may be relatively small and vice versa. For power frequencies, this relation is unproblematic. For higher frequencies however, the domain has to be made smaller and resultingly the resistivity higher. Eventually, the difference between the air domain resistivity and the contact domain resistivity will become small so that a non-negligible fraction of the current may flow in the air region. This is highly undesirable of course. At this point, only a high resolution mesh in combination with larger contact resistance domains can help. Generally, it is of course advisable to select the contact domain dimensions in such a way that the skin effect is negligible. The skin depth is determined approximately as follows:

$$\delta = \sqrt{\frac{2 \cdot \rho}{2 \cdot f \cdot \mu}} \tag{3.72}$$

with the skin depth δ, the resistivity ρ, the frequency f and the permeability μ which is a product of the permeability constant μ_0 and the relative magnetic permeability μ_r. For simplicity reasons, we assume the latter to be unity which does not influence the simulation.

Alternatively, it is also possible to mesh the contact domains so densely that even if the current should only be flowing on the outside, the correct resistance is still ensured. This leads to increased complexity and hence computation time however so this is not advisable.

Another possibility to model contact resistances reduces the required computational effort even further and also allows two-dimensional modelling. This approximation was developed by V. Zermeno (private communication). The influence of the contact resistances on the current distributions may be included in a boundary condition that is directly patched into the partial differential equation. A domain specific offset has to be added to the conservative flux term (compare with Eq. 3.46) and the respective offsets declared domain-wise:

$$\Gamma = \begin{bmatrix} 0 & E_z + V \\ -E_z + V & 0 \end{bmatrix}.$$

(3.73)

For the superconducting domains and the air domain, this offset is null. For the contact resistance domains, the offset V is declared as:

$$V = \sum_{i=1}^{n} R_i \cdot \int j_z dS_i$$

(3.74)

for each i up to n total domains. Strictly speaking, this approach is only possible for quasi-static simulations where E is virtually constant with time and uniform over a domain. Using the offset assumes average values of the electric field which is not the case when temporal variations are present. On the other hand, it is presumed that the approach is valid as a qualitative reliable approximation in the very low frequency regime as well since initial simulations showed plausible results.

3.2.7 Meshing

While in the case of choosing a mathematically convenient analytic function for μ_r speed was the only concern, the quality of discretisation also directly influences the accuracy of the simulation. Not only do the degrees of freedom depend directly on the size and distribution of the mesh, a good mesh also helps the simulation converge towards the solution more quickly. Conversely, a bad mesh might impede convergence altogether, though mathematically a solution exists. Even worse are erratic results which might not be immediately obvious. This is why numerical simulations require careful plausibility checks.

Instead of Lagrange (also called nodal) elements often used in FEM, the H-model employs curl-conforming Whitney 1 (also called Nédélec, edge- or curl-) elements. Theoretically, edge elements, which are computationally more expensive, are not required when simulating simple magnetic field models since nodal elements also work. However, the latter supposedly rely on the properties of the media, and consequently the electric and magnetic field, to be continuous functions of the spatial coordinates for mathematical rigidity [Mur94]. In geometries like the ones subsequently investigated, this is true only for the air region. Alternatively, a very fine mesh would also work, discretising the discontinuities. Consequently, a much larger number of degrees of freedom would be required to achieve the same accuracy as with edge elements.

Using nodal elements, the simulation could be considerably accelerated for the air regions. Unfortunately, we were not able to find suitable coupling constraints allowing us to connect domains with different types of elements (such as nodal and edge elements) in tests using our FEM software as the flux conservation could not be ensured. The reason this was investigated is that nodal elements are better suited for modelling the air domain and should result in faster convergence of the model.

It has been stated that edge elements inherently ensure divergence free fields [BGM07] but this is a misconception of the fact that edge elements themselves are internally free of divergence. Divergences may occur at the interface of two adjacent elements (only the continuity of the tangential field component is enforced) as the normal field component is free to jump at each of the interfaces [Mur94]. However, in our investigations we carefully verified that the solutions are correct and the possibly occurring local errors apparently do not influence the overall precision. The effect is mitigated in general in large parts by constructing high quality meshes.

High quality meshes comprise preferably isosceles triangular elements as the normal components of the state variable u to be solved of elongated elements may become almost degenerate with respect to each other, resulting in large local errors [Mur94]. Since the edge elements are linear internally, extremely elongated elements will lead to strong local dislocation artefacts, where regions of high contrast may show non-physical tendrils. Averaging shows these tendrils as smudges diluting high-contrast areas.

Also problematic are inverted elements, where quadrangular meshes in particular show re-entrant corners which lead to spurious local results. This need not necessarily lead to non-converging or wrong global solutions however. Due to peculiarities in the FEM software COMSOL, also triangular elements may be inverted.

Often, the perpendicular magnetic field component is the most important for coated conductors. Using quadrangular elements to model these geometries leads to the component of elevated interest being aligned with the normal component of the elements therefore enhancing the accuracy besides saving half of the elements and, consequently, half of the degrees of freedom (as opposed to using triangular elements) in those regions.

Manually constraining the mesh using mapped meshes and forcing certain distributions is usually preferable to using the automatic mesher for the complete geometry, see the comparison in Fig. 3.8.

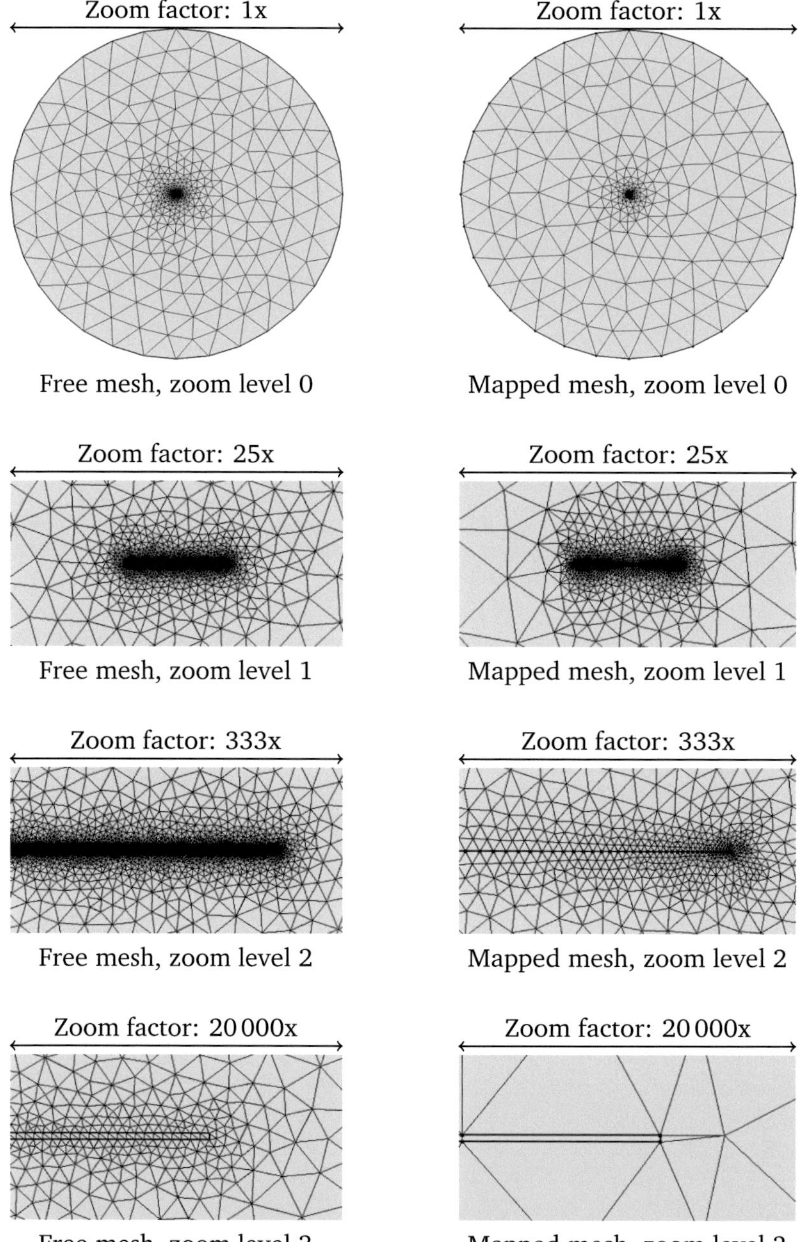

Figure 3.8: Comparison of free and mapped mesh. If an unconstrained automatic triangular mesh is used for a simple coated conductor, roughly 100 000 elements are needed. If a manually adjusted quadrangular mesh is used for the same geometry, only around 2 000 elements are required – without loss of precision. A reduction by a factor of 50.

4 Experimental Set-ups

To verify the numerical simulations with experimental investigations, two methods were employed for applied magnetic field measurements: the magnetisation loss and the calibration free method, see Cap. 4.2.1 and Cap. 4.2.2. The losses occurring under applied electric transport currents were measured using a lock-in amplifier and with digital acquisition units enabling high-speed measurements, see Cap. 4.3.1 and Cap 4.3.2. To discern the critical current and n_p-value of the coated conductor tapes, high speed, high precision four-point measurements were conducted, see Cap. 4.1.

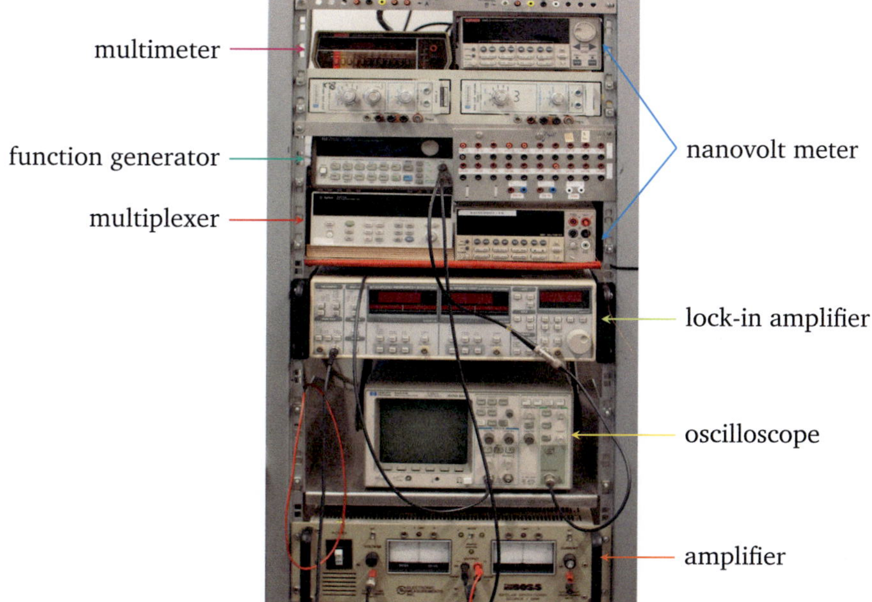

Figure 4.1: Picture of the instrumentation used for experiments. Not shown is a PC used for instrument control and data acquisition. The oscilloscope is used for in-situ surveillance to notice erroneous measurements in a timely manner. Also not shown here is the capacitor bank used for the calibration free method, see Cap. 4.2.2 as well as the parallel amplifier array used for the high-speed DAQ applied transport current measurements, see Cap. 4.3.2.

4.1 Critical Current Measurements

The difficulty in conducting critical current measurements on coated conductor tapes is to avoid destroying the tapes in the process. Due to their high n_p-value and thus fast transition from superconducting to ohmic behaviour and often little stabilisation, they are extremely delicate and burn out quickly. Since exact data on tape behaviour is required for realistic numerical simulations, the tapes have to be measured nevertheless and care has to be taken not to damage them. An additional complication is the fact that the tapes cannot be measured slowly as they dissipate heat close to their transition point, which leads to heating and thus falsifies the measurements. Ideally, the measurements are conducted as fast as possible while at the same time taking care not to burn out the coated conductors.

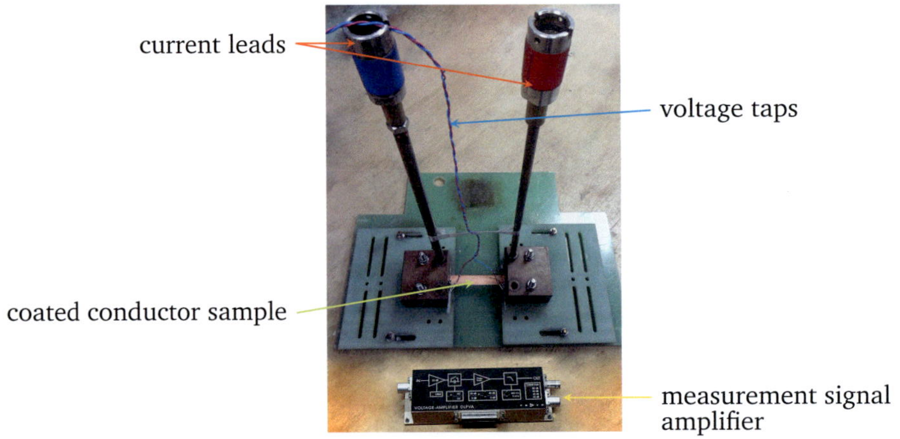

Figure 4.2: Picture of the critical current measuring system with one pair of voltage taps and one measurement signal amplifier shown. In the actual measurements, three pairs of voltage taps and measurement signal amplifiers are used with additional taps on the current leads and the shunt. The shunt (not shown for simplicity) is fastened to the empty mounting points on the copper terminals.

A special measuring system was set up that has extensive protective measures. Firstly, a large shunt with a small resistance was put in parallel to the coated conductor in order to carry the current once the coated conductor transitions to isolating behaviour. Secondly, the voltage was measured at the current terminals, on the shunt and on the coated conductor itself so a holistic monitoring was possible. Thirdly, high quality, high gain measurement signal amplifiers were used in order to register the transient behaviour as soon as possible. The zero-drift ($0.5\,\mu\mathrm{V\,K^{-1}}$) signal amplifiers have a very low input noise of 0.4 nV/Hz$^{0.5}$ and up to 100 dB amplification. Fourthly, high speed (16 bit, 250 kHz) as well as high precision (20 bit, 250 Hz) digital acquisition units were used concurrently so that not only very exact measurement points could be obtained, but the burn out protection could react very quickly. And lastly, the measuring speed was reduced to 0.5 A s^{-1} (down from 1 A s^{-1}) once the beginning of the transition from superconducting to ohmic behaviour was detected.

The shunt was selected such that its resistance is below the burn out threshold of the coated conductor. A copper cable of 1 cm^2 diameter was patched in parallel to the sample and the contacts were polished beforehand to ensure a low contact resistance. The copper cable was about 15 cm long. Such a cable has a total resistance of somewhere between 20 $\mu\Omega$ and 40 $\mu\Omega$, with contact resistances roughly 50 $\mu\Omega$ at most. Everything below 30 $\mu\Omega$ cm^{-1} was tested to be safe for a coated conductor. The threshold of the voltage detected on the taps on the superconductor above which the current source was shut down was set to 10 $\mu\Omega$. For an exemplary measurement, see Fig. 2.6. With the gain of the amplifiers set to 40 dB, this means voltages somewhere in the order of 1 nΩ to 10 nΩ can be detected reliably in the 0.5 s period in between steps of the current ramp. With the voltage taps placed 5 cm apart and the critical current criterion being 1 μV cm^{-1} (see Cap. 2.5), this leaves a more than ample safety margin while at the same time insuring precise data acquisition.

4.2 Background Field Magnetisation Loss Measurements

Background fields of a high enough amplitude will cause flux penetration into and magnetisation of the superconducting sample. Due to the mechanism outlined in Cap. 2.6.1, a change in magnetisation will necessarily lead to energy dissipation. There are two complementary methods for measuring the dissipated energy: as a temperature difference (see Cap. 4.2.1) and measuring the energy loss of the excitatory magnetic field (see Cap. 4.2.1 and 4.2.2).

4.2.1 Combined Thermal/Complex Susceptibility Measurement Method

Calorimetric methods have the advantage of simplicity. The measurements of temperature differences or gas flow due to cryogenic coolant boil-off and the resulting correlation with the dissipated power are not afflicted with complex calibrations and the resulting errors. They are hindered however by their long time constants, especially at liquid Nitrogen and even at liquid Helium temperatures. Calorimetric measurements are therefore either slow or inexact [Sch00].

The combined thermal/complex susceptibility measuring apparatus used in the investigations contains two distinct measuring systems to combine the advantages mentioned above and measures both calorimetrically as well as magnetically. For an exemplary measurement see Fig. 4.3, which shows the temperature rise due to applied magnetic background field, the resulting hysteretic losses and the subsequent replication of this temperature rise using a resistive heating circuit. For a schematic view of the measuring system, see Fig. 4.4. The calorimetric method registers the temperature and the rise in said observable due to applied oscillating magnetic fields causing dissipation. Later, this temperature difference is matched by resistively heating the sample with known power.

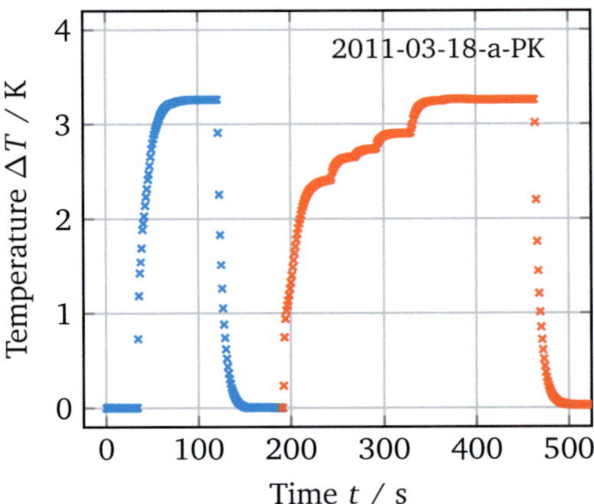

Figure 4.3: Exemplary calorimetric measurement; the temperature rise induced by the hysteretic loss is plotted in blue while the orange plot shows the temperature rise due to the heater and the stepping up of the current in order to match the temperature rise due to the hysteretic loss.

With the copper resistance changing by about 2.9 % per 1 K around 77 K, the change in temperature is:

$$\Delta T = \frac{\left(\frac{R_{T,in}}{R_{T,out}} - 1\right)}{2.9\,\%\,\mathrm{K}^{-1}}. \tag{4.1}$$

In order to acquire less noisy data, the sample, heater and sensors are enclosed in a vessel with thermal insulation in order to have a thermal resistance between sample and cryogenic coolant bath. $R_{T,in}$ and $R_{T,out}$ are the temperature sensors inside and outside of the thermal insulation around the sample. This enables steady-state measurements after thermal equilibrium has been achieved. For an exemplary measurement, see Fig. 4.3. The power dissipated by the resistive heater is calculated from the voltage U_h and current I_h measured in the heater circuit:

$$P_{dc} = U_h \cdot I_h.$$

(4.2)

Lastly, the applied magnetic field needs to be recorded which is calculated from the voltage in the pick-up coil U_{pc}, the frequency f and a calibration constant of the pick-up coil c_p:

$$B_a = \frac{U_{pc} \cdot c_p}{f}.$$

(4.3)

The calibration constant of the pick-up coil c_p is determined by leveraging Faraday's law:

$$U = N \cdot \frac{B_a \cdot S}{t}$$

(4.4)

which, rearranged, leads to:

$$c_p = (N \cdot S)^{-1} = \frac{B_a \cdot f}{U}$$

(4.5)

with the number of coil turns N and the surface S which is the cross-section of the pick-up coil. For a narrow temperature range, the response of the system may be assumed linear, so that using a thermal constant c_t, dependent on the thermal conductivity of the insulation and geometry is defined as:

$$c_t = \frac{P_{dc}}{dT_{dc}}$$

(4.6)

and may be used to calculate the power dissipation:

$$P_{ac} = c_t \cdot dT_{ac}.$$

(4.7)

In order to obtain precise measurements, dT_{ac} and dT_{dc} should be as close as possible.

While measuring the difference in temperature is relatively simple, the method is slow since it relies on establishing thermal equilibrium. Magnetisation measurements are much faster, a calibration is required for every individual sample however. The calib-

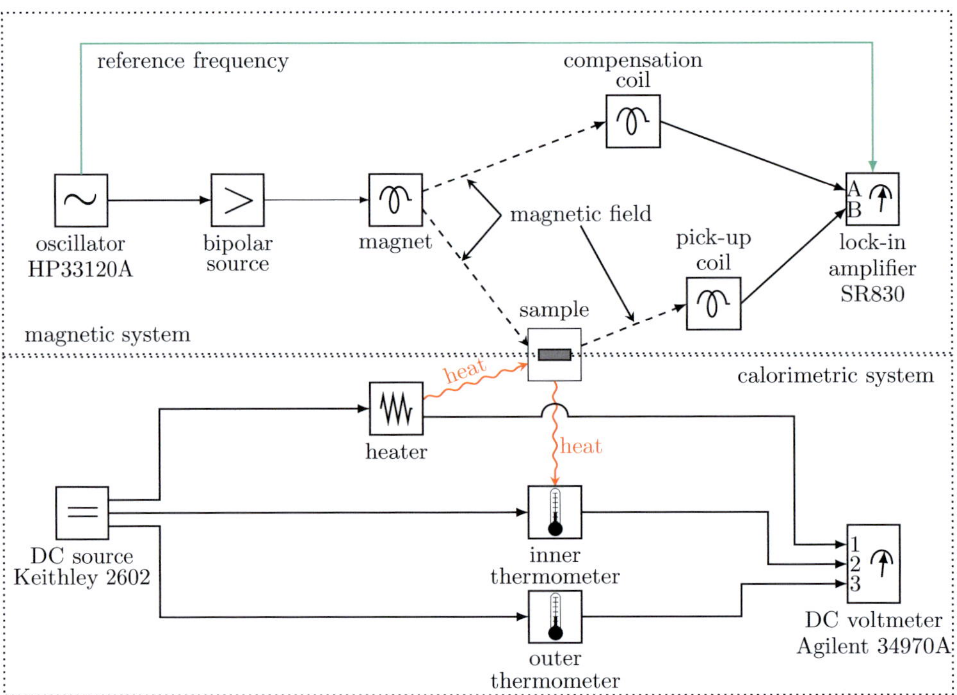

Figure 4.4: Block diagram of magnetisation loss measuring setup with included calorimetric calibration and measuring system.

ration constant is obtained using mixed calorimetric/magnetisation measurements. When using the magnetisation measurement system, two coils are used to register the measuring signal: the pick-up coil in the sample holder and the compensation coil situated close to the outer coils producing the magnetic field. The two measuring coils are connected in series in such a way that the total signal is zero without a sample (anti-inductive configuration). Additionally, the signal of the second coil, the compensation coil, is also registered isolatedly, meaning not in series with the first measuring coil. This is done in order to measure the applied flux density B_a.

In practise the compensated signal deviates from zero due to the pick-up and the compensation coil not being ideally tuned. This does not pose a problem because the integration of the voltage over one cycle still yields zero [Sch00]. When a sample is present, the voltage is proportional to the time derivative of the sample magnetisation:

$$U(t) \propto \frac{d\Phi(t)}{dt}. \tag{4.8}$$

Where the magnetic flux Φ is defined as an integral over the surface S:

$$\Phi = \int_S \vec{B} dS. \tag{4.9}$$

Since the signals are recorded using a lock-in amplifier, two loss components of the voltage induced in the coils can be measured: in-phase with the current driving the magnetisation coils and shifted by $90°$ out-of-phase. The former represents the loss component while the latter is the inductive component, proportional to the magnetic energy stored in the coils.

The calibration factor c_f required for the magnetisation measurements may be determined using Eq. 4.7:

$$c_f = \frac{P_{ac}}{B_a \cdot U_{p,y}}. \tag{4.10}$$

For Eq. 4.10, the inductive voltage component registered in the pick-up coil $U_{p,x}$ is not of interest. Only the loss voltage component $U_{p,y}$ is required. The inductive voltage component of the pick-up coil is in-phase with the voltage registered in the compensation coil U_{comp} which is why pick-up coil and compensation coil need to be tuned to supply the same signal amplitude (ideally). By subtracting the compensation coil voltage signal U_{comp} from the pick-up coil voltage signal and thus attenuating the stray signal of the inductive component it is possible to register the loss voltage component with much higher precision.

The ac loss dissipated in the system is then:

$$Q = \frac{B_a \cdot U_{p,y} \cdot c_f}{f}. \tag{4.11}$$

4.2.2 Calibration Free Method

When a sample is subjected to an ac field generated by a coil system, the power dissipated by the superconductor has to be supplied by the current source and is a fraction of the ac power supplied to the coil system [Voj10, pp. 54 ff.]. If it were possible to distinguish this fraction from all the other sources of loss in the system, the absolute value of the ac loss could be obtained. The biggest influence will be the loss of the coil system. Using two identical coil systems having been connected in series and using two pick-up coils connected in series but with reversed polarity results in the signals of the pick-up coils inside the two coil systems compensating each other, see Fig. 4.5.

The variable capacitor shown in parallel with the coil system is required when driving the system at high magnetic fields. The current source is not used to directly drive the coils but only pumps energy into the resonant circuit. Since the capacity has to be tuned for every frequency, we use a capacitor bank with disconnectable capacitors. The tuning is accomplished by simply having the current source drive the coils and connecting and disconnecting capacitors until an also connected Rogowski coil registers maximal electric current flowing through the system.

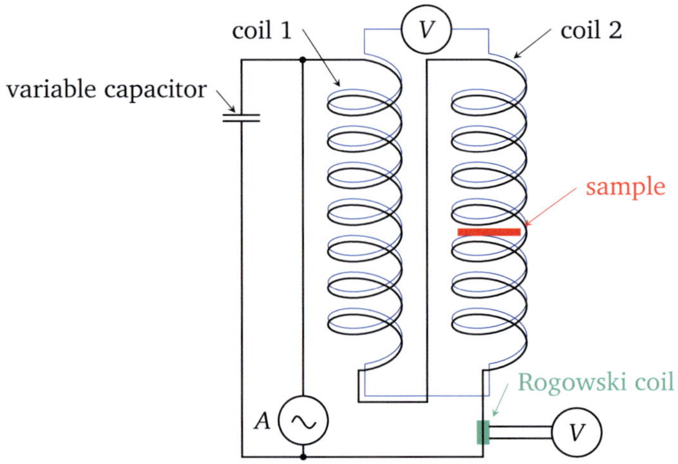

Figure 4.5: Calibration free magnetisation loss measurement system; note that both the voltages of the pick-up coils as well as of the Rogowski coil are measured using a lock-in amplifier. The variable capacity is tunable in order to reach maximal magnetic field strength in the coil system while not damaging the current source. The resonant frequency depends on the frequency of the applied background field and on the current source, cables and environmental influences like temperature or magnetic surroundings.

Supposing the two coil systems are identical then the only measurable signal will be that due to the sample being present in one of them. The measured voltage will be [vGV05]:

$$U = \frac{d\Phi}{dt}.$$

(4.12)

The signal should be zero without a sample present and only carry information about the dissipated power. The electrical current powering the magnet coils is measured with a Rogowski coil. Theoretically, voltage taps on the primary circuit powering the coils are possible. However, this method of electrical current measurement is not advisable due to the risk of destroying the lock-in amplifier through overdriving with too high signals.

As with the magnetisation method, the registered voltage has a component in phase with the electric current driving the magnet coils and an out-of-phase component. Using the lock-in amplifier, the in-phase component is measured and the hysteretic power loss P is:

$$P = I \cdot \mathrm{Re}(U). \tag{4.13}$$

Due to imperfections in the coil system, the signal is not zero even without a sample being present however. In order to correct for these spurious signals, the response of the coils has to be tuned in both the resistive loss as well as the inductive voltage component. The resistive component is most easily influenced by introducing a conducting material such as a metallic block into the coil with the lower losses, thereby increasing the losses and thus balancing the signals of the two pick-up coils. The inductive voltage component is corrected by inserting a compensation coil into the system and connecting it in series with the two pick-up coils. Both the metal as well as the compensation coil have to be positioned precisely in such a way as to correct the signal. Since the response of the system is dependent on frequency and conducting materials in its surrounding, the compensation should be checked before each measurement.

The compensation free system was fitted with a few improvements to speed up the measuring process: before measuring each data point, the sensitivity has to be adjusted. In order not to overdrive the lock-in, the highest setting is always used in the first step. The system is then used to simply evaluate whether the signal was between 50 % and 100 % and stopped increasing the sensitivity if that was the case. Else, the sensitivity would be increased and the procedure repeated. Each step takes about ten seconds so the whole process takes a lot of time.

An algorithm was developed that estimates the correct sensitivity and jumps up the resistivity accordingly in only two or three steps so that the process takes much less time, speeding up the whole measuring cycle.

Additionally, instead of moving a metallic block about in the liquid cryogen bath which is tedious, a tunable coil with adjustable resistance was installed. Another improvement regards the distribution of measuring points: in order to record linear or logarithmic sweeps, loops or focus a certain region while sweeping, an additional program allows generating complex measurement configurations.

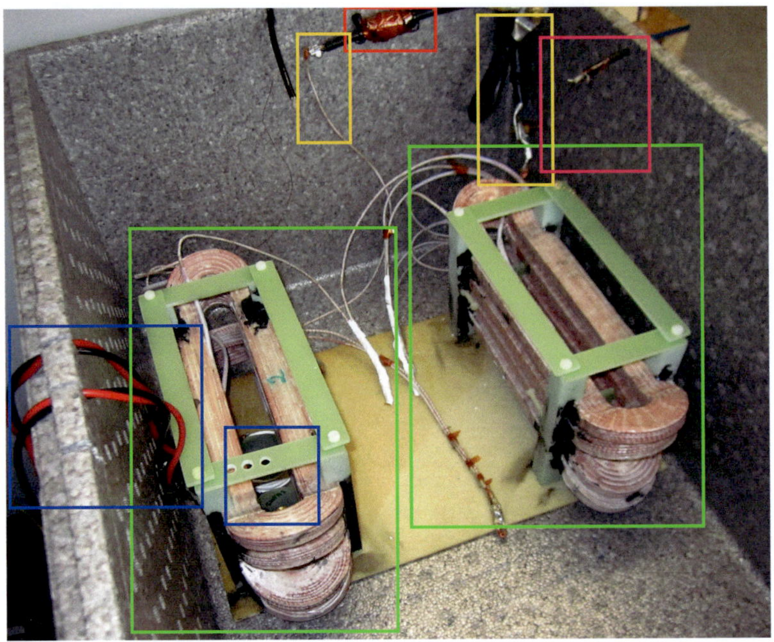

Figure 4.6: Picture of the calibration free method coil system. Shown are the two coil systems (green) with the current leads feeding the coils (yellow), the Rogowski coil used for current measurements (red), the contacts for the pick-up coil wires (pink) and the tunable compensation coils with contacts (blue).

4.3 Transport Current Loss Measurements

Multiple methods to measure transport current losses in superconducting components exist. The thermal and boil-off methods use the rise in temperature and the subsequent evaporation of cryogenic liquids respectively to measure the energy dissipation in superconducting components. The secondary voltage method measures the energy delivered to the sample by the source instead of the dissipated energy [vG02]. One of the most convenient approaches is to simply use voltage taps and either use a lock-in amplifier (see. Cap. 4.3.1) or high-speed digital acquisition units (see. Cap. 4.3.2).

4.3.1 Lock-in Measurements

Using the lock-in measurement technique for applied electrical transport current measurements is conceptually very similar to the magnetisation measurements. The alternating electric transport current that is applied to the sample leads to hysteretic losses in the superconducting tape. If the transport current has not reached I_c however, a certain region in the centre of the superconductor will be void of current. For both the lock-in as well as the high-speed digital acquisition methods, care has to be taken when placing and connecting the voltage taps, since the hysteretic losses can not be measured by only detecting the voltage drop over the coated conductor.

The magnetic field surrounding the conductor is equally important which is why the wires connecting the voltage taps have to be laid out in such a way that an area as large as possible is covered. On the other hand, a larger loop will have a higher noise signal. The signal is proportional to the voltage drop and the time derivative of the magnetic flux. With constrictions of the measuring system and practical considerations in mind, previous research suggests that an area at least three times the half width of the tape is required for a satisfactory accuracy of more than 95 % (see Fig. 4.7) [YHB+96]. Increasing the area of the loop helps to record more of the magnetic field and therefore improves accuracy of course.

Placing the voltage taps in the central part of the tape entails an underestimation of the actual losses that is gradually improved by enlarging the area of the loop. The voltage taps may also be placed on the edge on one side of the tape. This leads to an overestimation of the losses that is also gradually improved by enlarging the area of the loop.

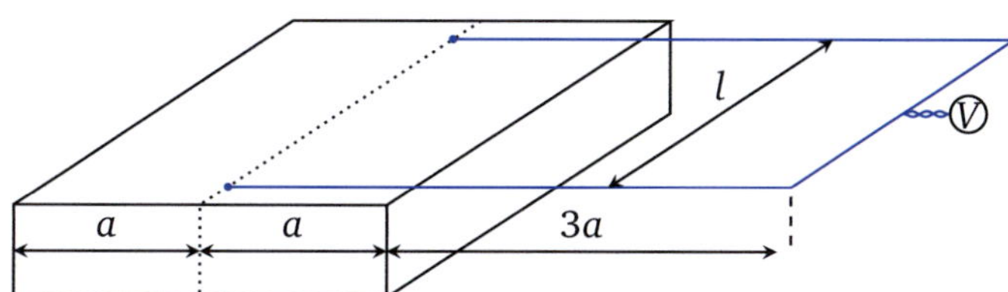

Figure 4.7: Schematical drawing of where and how to place voltage taps when conducting electric transport current measurements. The hysteretic losses cannot be measured by only detecting the voltage drop over the coated conductor. The magnetic field surrounding the conductor is equally important which is why the wires connecting the voltage taps have to be laid out in such a way that an area as large as possible is covered. Previous research suggests that an area at least three times the half width of the tape is required [YHB+96].

Due to the non-linear behaviour of the superconductor during flux penetration, even when a harmonic electric transport current is applied, the resulting voltage signal is not harmonic. The lock-in amplifier is therefore used to obtain the voltage component in-phase with the applied electric transport current. The transport loss is equal to the product of the root mean square of the transport current and the real part of the root mean square voltage:

$$P = I_{\mathrm{rms}} \cdot \mathrm{Re}(U_{\mathrm{rms}}). \tag{4.14}$$

Because the loss is frequency independent when it is purely hysteretic in origin, considering the energy dissipation per cycle Q may prove useful:

$$Q = \frac{I_{\mathrm{rms}} \cdot \mathrm{Re}(U_{\mathrm{rms}})}{f}. \tag{4.15}$$

In the applied transport current measurements using a lock-in amplifier, a Rogowski coil is used to acquire the electric current signal just like in the applied background magnetic field measurements (see Fig. 4.8). The current transducer also shown is only used for high-speed measurements (see Cap. 4.3.2).

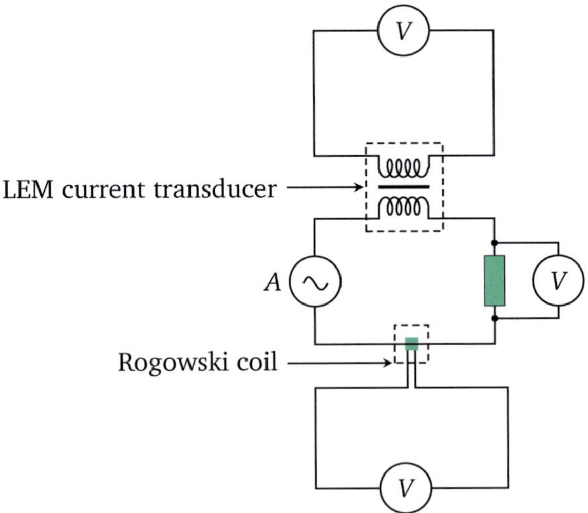

Figure 4.8: System for electrical transport current loss measurements with both a current transducer as well as a Rogowski coil for reference electrical current measurements.

4.3.2 High-speed Digital Acquisition Measurements

Instead of using a lock-in amplifier for signal recording, using high-speed digital acquisition units is also possible. Precise voltage and current measurements with a sample rate of up to 500 kHz provide data that can be used to directly calculate the instantaneous losses:

$$P(t) = I(t) \cdot U(t). \tag{4.16}$$

Q is then:

$$Q = \int_0^{1/f} P(t)dt = \int_0^{1/f} I(t) \cdot U(t)dt. \tag{4.17}$$

Using high-speed digital acquisition units has the advantage of speed: since phase alignment does not have to be enforced and monitored, not only is one error source removed, but also one setup step. In addition to using high-speed digital acquisition units, a current transducer was used to acquire the current signal. While at very high frequencies, its accuracy is limited by the time resolution, for power frequencies, it is more than sufficient. It has the advantage of not shifting the current signal by 90° like the Rogowski coil. The current transducer also does not have to be calibrated. In order to corroborate the high-speed measurements, reference data points were always taken using the slower lock-in method. These two systems can also be combined into one, see Fig. 4.8.

4.4 Comparison of Experimental Methods

Four different methods were presented in order to measure hysteretic losses in superconductors: two for applied electric transport currents and two for background magnetic field measurements. Conceptually, the calibration free method for measuring the hysteretic loss in background magnetic fields and the lock-in method for measuring with transport current are similar. Both use a lock-in amplifier in order to measure the real part of the voltage signal which records the superconductor's response in order to calculate the hysteretic loss. The post processing of the high-speed measurements is comparatively simple as the high temporal resolution allows the direct calculation of the losses from the current and voltage signals. The combined method is the most tedious and the slowest as a separate calibration has to be taken for every frequency and both the calibration as well as the following lock-in measurements take a long time. Refer to Tab. 4.1 for a quick overview.

Comparison of experimental measurement set-ups			
Method	Complexity	Speed	Accuracy
Combined Method	0	0	+
Calibration Free	+	+	0
Lock-in	+	0	+
HiSpeed DAQ	0	+	0
Evaporation	0	−	+

Table 4.1: Experimental measurement set-ups are compared with regards to the complexity of setting up the systems, the speed of the measurements and their accuracy.

5 Coated Conductors

Besides bulk specimen of *REBCO* that are being used as permanent magnets in specialised applications, the most basic entity of all superconducting appliances is the superconducting wire or tape. Cables and coils are wound from multiple such conductors, and most superconducting assemblies are in turn built from multiple cables or coils. In order to understand and predict the behaviour of even complex machinery, understanding the basic properties of cables and coils is fundamental.

There are two distinct coated conductor designs being manufactured: the specimen with non-magnetic substrates and a complex structure of buffer layers (see Cap. 2.7.3) and the RABiTS kind with ferromagnetic substrates (see Cap. 2.8.2). Since the ferromagnetic substrates have a rolling texture with a lattice parameter very close to that of *REBCO*, less buffer layers are required in order to provide a well defined surface on which to grow the superconducting layer. Coated conductors of the first kind were measured and simulated and the results will be discussed in Cap. 5.1 while the latter are discussed in Cap. 5.2.

5.1 Single Coated Conductors

Since the buffer layers are usually insulating and the conductivity of the stainless steel substrate is negligible, the electric and magnetic influence of both substrate as well as the buffer layers is disregarded when simulating. Only the superconducting layer is actually simulated electrodynamically. This is a valid approach since the steel substrate has a resistivity orders of magnitude smaller than the superconducting layer, there is an isolating barrier between them and the steel substrate is non-magnetic (unless RABiTS substrates are considered, see Cap. 5.2). Geometrically, the proportions are retained, see Fig. 5.1. This helps keep the complexity of the simulation down leading to reduced computational effort while not degrading simulation quality.

While the computational investigations were mostly carried out numerically, analytical formulations are often used as extreme test cases as mentioned in Cap. 3.1.3. In the following sections, reference is often made to these solutions. In order to reach an even higher degree of confidence in the results, the compliance of various numerical models is tested in Cap. 5.1.1. Subsequently, the influence of different parameters

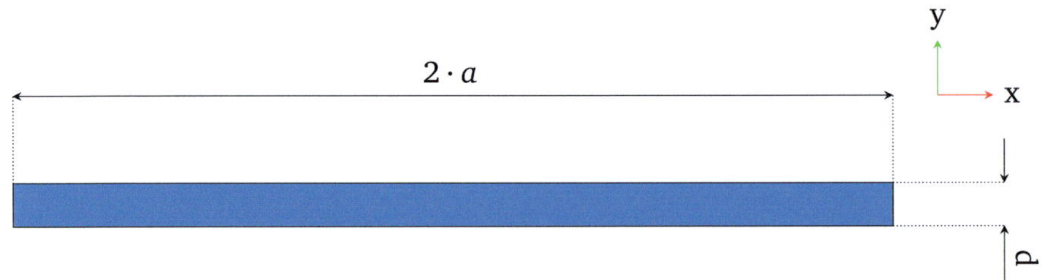

Figure 5.1: Drawing of a simulated coated conductor. The width, here denoted as $2a$ in reference to the formulas in Cap. 3, is usually 1 cm while the thickness d is usually 1 μm.

is investigated with regard to the effects on losses and magnetic field and electric current distributions. See Fig. 6.9 for an exemplary numerical simulation showing the relative magnetic field amplitude and direction surrounding the edge of a coated conductor with applied transport current.

5.1.1 Compliance of Numerical Models

In [GRK11], we were able to show that even using rough estimations like constant critical currents, reasonable agreement may be achieved between simulations and measurements when comparing hysteretic losses. And, unlike in the case of ferromagnetic substrates, no distinct features of the curves are lost. It is however important to understand the simplest case of just one superconducting layer in order to be able to understand more complex geometries. Mixed boundary conditions like applied electric transport current and applied background magnetic field for example yield quite a complex superconductor response.

In the case just mentioned, the behaviour may be understood more easily if the response is regarded as a superposition of both an applied electric transport current as well as a background magnetic field. It may also be computed as such (see Cap. 3.1.3 and Ref. [BI93]). While the work mentioned in Ref. [BI93] solves this problem analytically, it is still interesting to simulate this case: the analytic solution just mentioned only shows the magnetic field and electric current distributions at maximum penetration while all intermittent steps are not accessible. But this evolution from a virgin state of the electric current and magnetic field being zero everywhere in the superconductor is helpful in understanding the resulting complex current and field profiles.

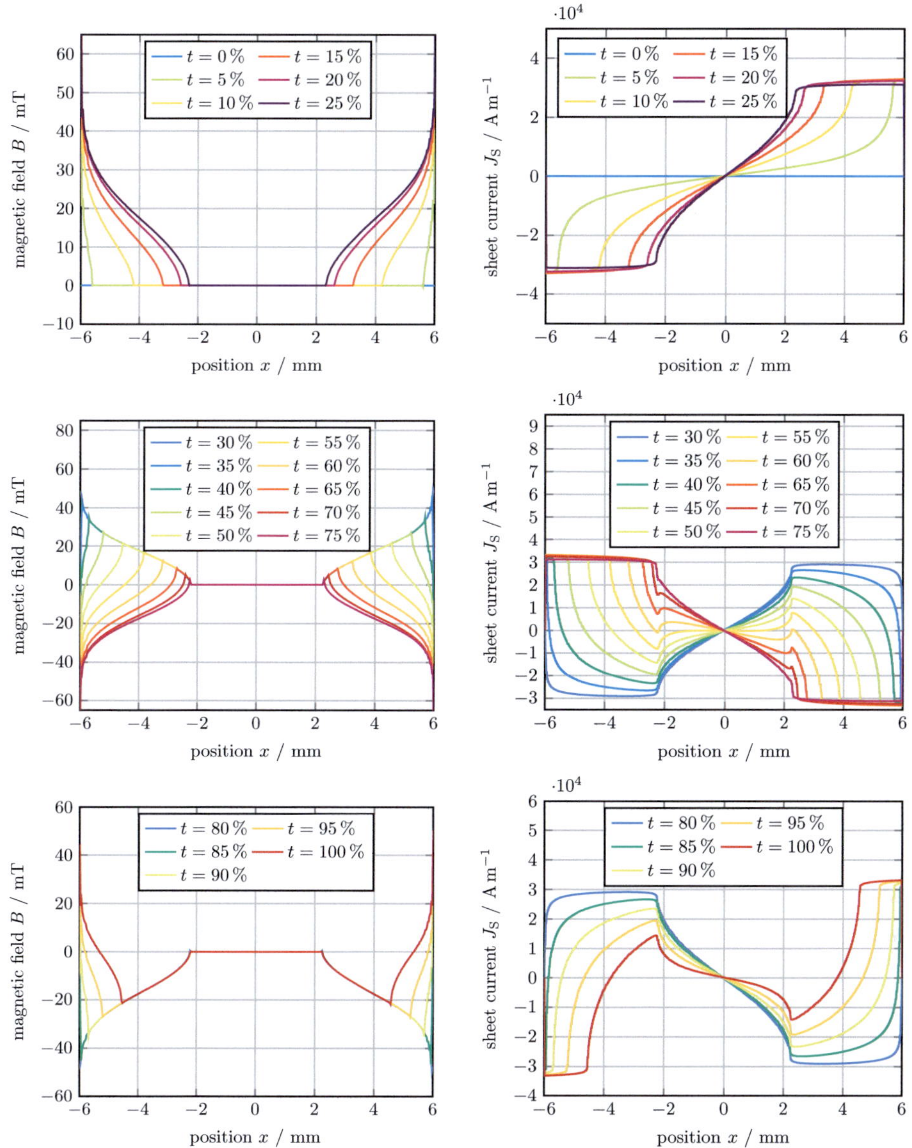

Figure 5.2: Temporal evolution of current and field profiles for applied background magnetic field in a superconducting strip of $1\,\mathrm{cm}$ width and a thickness of $1\,\mu\mathrm{m}$. The critical current is $300\,\mathrm{A}$ and the n_p-value 35, the time step is indicated as a percentage of a full cycle which in the case at hand is $0.02\,\mathrm{s}$ since the frequency of the background magnetic field is $50\,\mathrm{Hz}$.

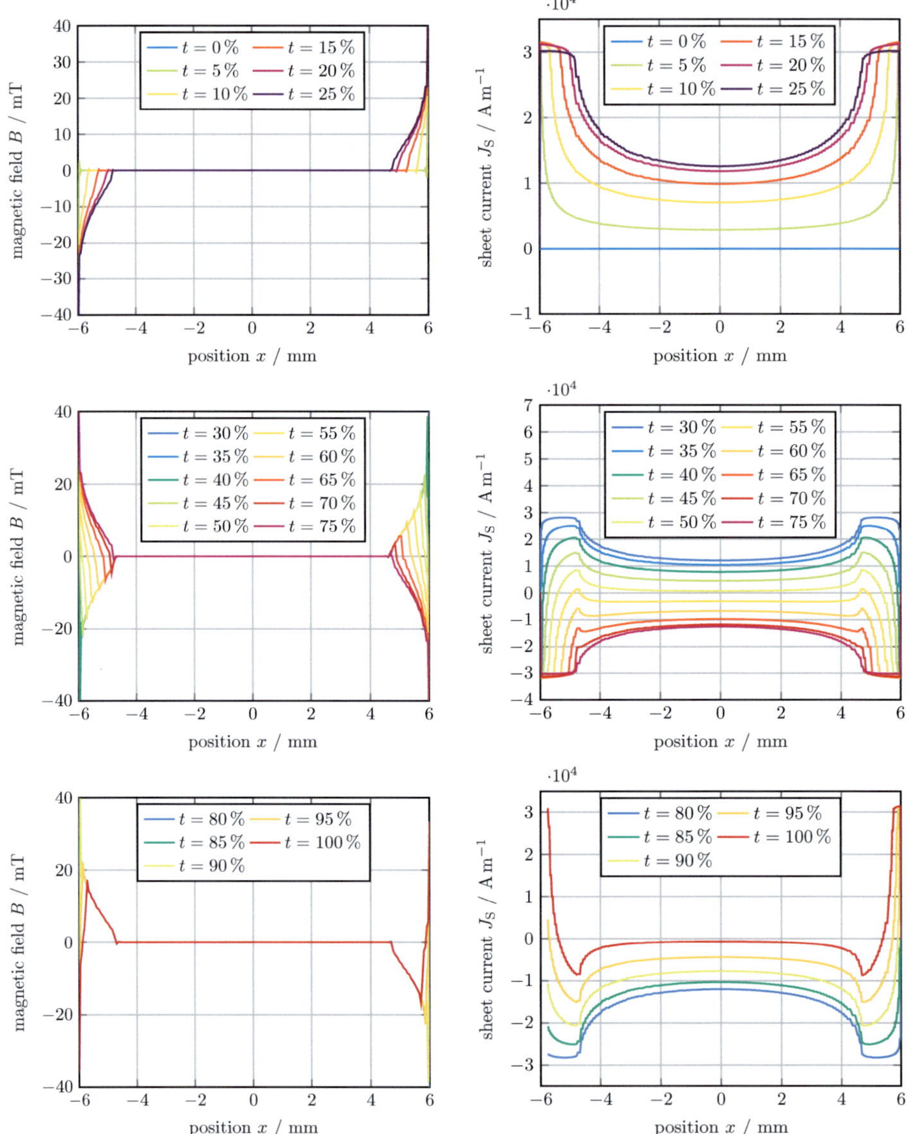

Figure 5.3: Temporal evolution of current and field profiles for applied electric transport current in a superconducting strip of $1\,\mathrm{cm}$ width and a thickness of $1\,\mu\mathrm{m}$. The critical current is $300\,\mathrm{A}$ and the n_p-value is 35, the time step is indicated as a percentage of a full cycle which in the case at hand is $0.02\,\mathrm{s}$ since the frequency of the background magnetic field is $50\,\mathrm{Hz}$.

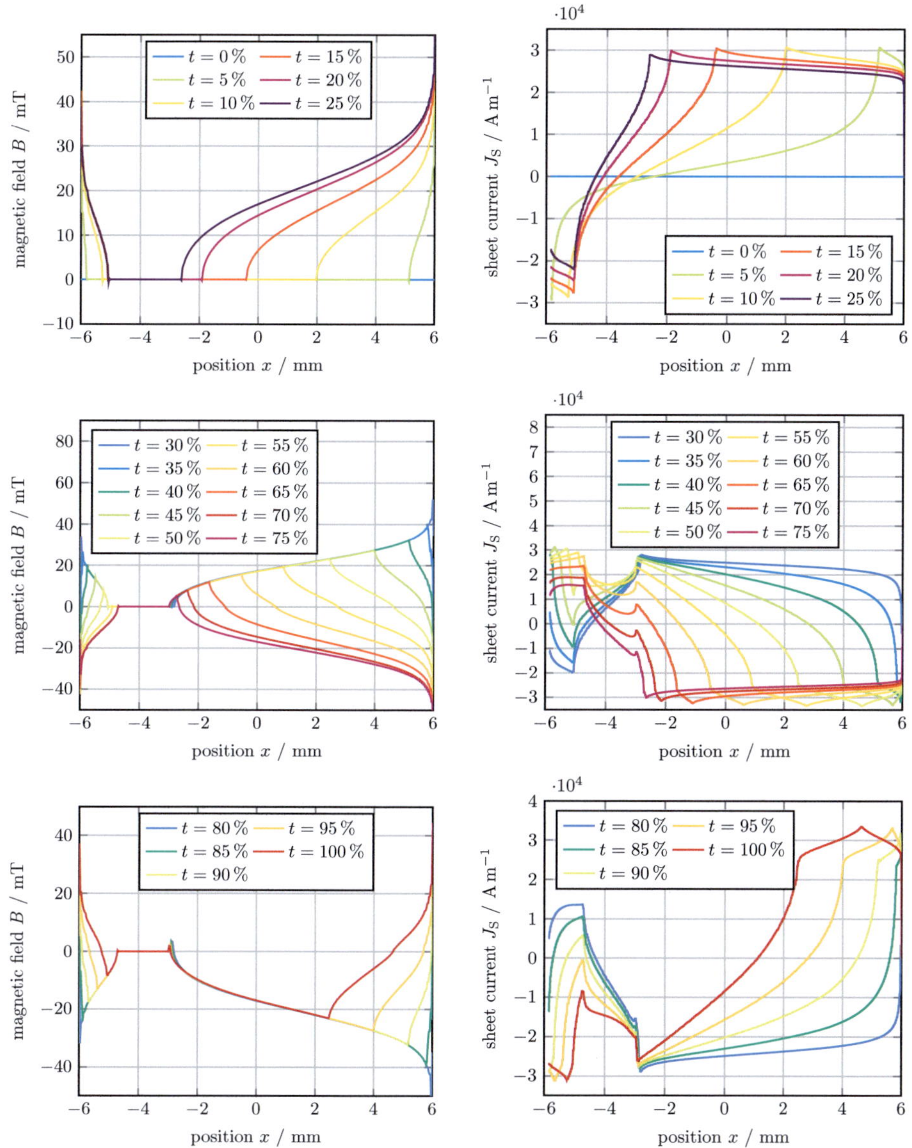

Figure 5.4: Temporal evolution of current and field profiles for applied electric transport current and background magnetic field in a superconducting strip of 1 cm width and a thickness of 1 μm. The critical current is 300 A, the n_p-value 35 and the time step is indicated as a percentage of a full cycle (here: 0.02 s) since the frequency of the background magnetic field is 50 Hz.

In order to better understand the evolution of field and current profiles, both temporally and depending on the models and parameters chosen, regard Fig. 5.2, Fig. 5.3 and Fig. 5.4. There are two possible simple load scenarios: applied background magnetic field (Fig. 5.2) and applied electric transport current (Fig. 5.3). These lead to complementary symmetric and antisymmetric profiles. The third possible scenario includes both applied electric transport current and applied background magnetic field concurrently, leading to more complicated profiles (Fig. 5.4).

Numerical models are required when excitations different from ramps or alternating current cycles need to be simulated, for example current spikes or disturbances.

The fact that for the combined load case the profiles are not symmetrical anymore with respect to the centre of the strip has bearing on how the losses evolve, especially when ferromagnetic materials are involved. This is why even for a relatively simple geometry such as a single strip, it is not easy to intuitively understand hysteretic losses. It is even more complicated for composite geometries. Due to the complex response of superconductors, every geometry has to be optimised for specific load scenarios.

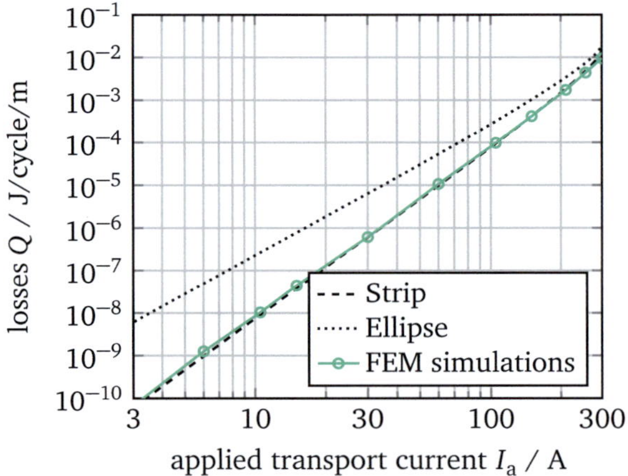

Figure 5.5: Hysteretic losses as a function of applied transport current in coated conductors. For low applied currents, the losses scale as I^4 in both the analytic solution (after [Nor70]) for the strip as well as the numerical simulation for a tape 12 mm wide and 1 μm thick with an I_c of 300 A and an n_p-value of 35. While the analytic solution assumes a one-dimensional strip, it is obvious that even though the coated conductor has a finite thickness, the solution for the strip predicts the behaviour very well. This also pertains to simulations utilising more than one mesh element across the width of the tape.

We compared simulation and analytic solution and found that due to the high aspect ratio of at least around $1 : 1000$ but up to $1 : 12000$, the hysteretic loss response of the coated conductors is virtually indistinguishable of the analytic solution. Even the assumption of constant critical current does not influence the results greatly as the application of a vertical magnetic field and the fact that *REBCO* conductors are usually more sensitive to perpendicular than parallel field with regard to the a-b-plane both mitigate the inaccuracies of this approximation (see Fig. 5.5). When using FEM simulations, the superconducting strip is often modelled by using only one mesh element across the width, reducing computational effort considerably by effectively suppressing lateral electric currents due to averaging and only allowing longitudinal currents to flow. It is important to bear this in mind when analysing configurations with incident background magnetic fields of other than perpendicular orientation.

If the dependence of the critical current density on the magnetic field is important and if not either applied electric transport currents (where the field orientation is less important) or applied magnetic fields perpendicular to the superconductor are investigated, it is advisable to adapt the model as the anisotropy of the superconductor has to be considered. This includes both meshing the superconductor with more than one element across its thickness and taking the critical current density dependence into account.

5.1.2 Influence of Power-Law Exponent

It has been mentioned that the analytic solution is only able to reproduce cases where $n_p = \infty$. This is due to the fact that Brandt [BI93] assumes a critical state model (see Cap. 3.1.1) whereas the H-model is based on the power-law (see Cap. 2.5). The differences are quite pronounced, as can be seen in Fig. 5.6. The analytic solution is an edge case of the H-model, with n_p tending to infinity. The other extremal parameter is the frequency f, which will be discussed in Cap. 5.1.3.

The most striking feature of the comparison is that at low n_p-values, an overcritical sheet current J_S may flow. The higher n_p becomes, the less pronounced this effect and the closer the resulting electric current and magnetic field profiles are to the analytic case. Solving the H-model – based on the power-law – for $n_p = \infty$ is not possible mathematically which is why the profile shown was obtained using the adaptive-resistive method and is merely shown for reference.

The adaptive-resistivity method [GH05, FFG10] uses iterative adjustment of the material resistivity in order to approach the critical state model. Initially, a constant resistivity $\rho = \rho_0$ is assumed in all $i = 1, ..., M$ elements of the superconducting domains. In a subsequent iterative process, the resistivity of all the elements is adjusted according to:

$$\rho_i^{k+1} = \max\left(\rho_i^k \frac{|j_i^k|}{j_c}, \rho_\epsilon \right) \tag{5.1}$$

where k is a time step counter and j and j_c are the current density and the critical current density, respectively. The minimum meaningful value for the resistivity is given by ρ_ϵ. This process continues until:

$$\frac{\rho_i^{k+1} - \rho_i^k}{\rho_i^k} < \epsilon \tag{5.2}$$

with the positive infinitesimal ϵ determining the simulation accuracy.

The power-law exponent n_p governs the abruptness of the transition between super-conducting and normal-conducting state. The transition becomes steeper with a rising value of n_p. This explains the observed behaviour of currents above critical sheet

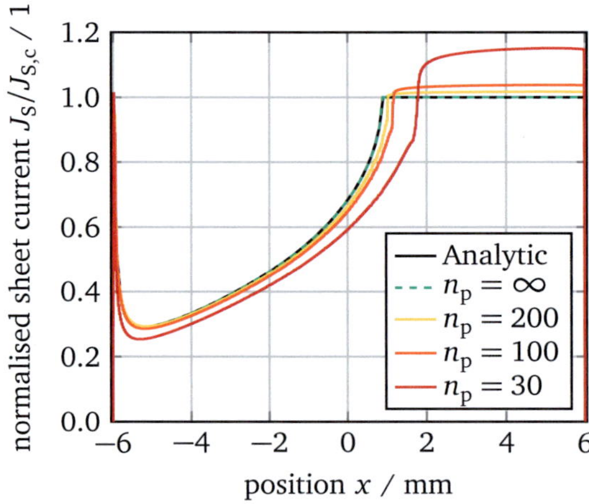

Figure 5.6: Influence of the power-law exponent n_p; shown are the results of a 12 mm wide and 1 μm thick coated conductor with a critical electric current I_c of 300 A and a critical magnetic field of $H_c = J_{S,c} \cdot d/\pi$ where d is the thickness of the conductor. The tape was loaded with an applied electric transport current of $0.7 \cdot I_c$ and a background magnetic field of $0.7 \cdot H_c$. The numerical result with infinite power-law exponent n_p was simulated using the adaptive resistivity model. The other three current profiles are the result of H-model simulations.

current $J_{S,c}$ flowing: the superconducting state does not collapse abruptly but rather fades away smoothly. Overcritical currents are not steady-state effects however, which is why the frequency influences superconducting behaviour ever more strongly the lower the value of power-law exponent n_p becomes. For a discussion of this frequency influence, refer to Cap. 5.1.3.

5.1.3 Frequency Influence on the Electric Current Profiles

In the critical state model, the frequency does not have any influence on the energy dissipation since both the magnetic field and electric current distributions as well as the hysteretic losses themselves are frequency independent. This means that while the hysteretic losses per time rise as the frequency increases, the losses per cycle do not.

As previously mentioned however (see Cap. 5.1.2), when using the power law and low n_p-values in the H-model, the electric current distribution is not in equilibrium anymore. Looking at Fig. 5.7, the effect of ever decreasing frequency is obvious: towards infinitely small frequencies or, equivalently, constant loads, the initially well defined current distribution dilutes and, given enough time, the current in a superconducting strip will tend to distribute uniformly towards infinity: while the total current flowing in the superconducting strip remains constant, the current distribution may change substantially. In this case, the frequency does have direct influence on the energy dissipation as the hysteretic losses depend non-linearly on the current distribution.

5.2 Coated Conductor Bilayer with Ferromagnetic Substrate

Rolling certain metallic alloys results in emerging textures. These are called called rolling textures and are due to specific crystal structures. Special physical properties result from these textures. For the previously mentioned $Ni_{5\%}W_{95\%}$ alloy (see Cap. 2.8.2), this texture is extremely conducive to the growing of *REBCO* superconducting layers since the rolling texture and the superconducting layer's crystal unit cell and orientation are very similar [GNB+96]. Using the RABiTS approach, deposition of superconducting layers on metallic substrates is possible with few or even no buffer layers present [GNB+96, GNK+97]. However, the ferromagnetic properties of the substrate have to be considered, because not only do they influence the electric current and magnetic field profiles and the hysteretic loss profile of the superconducting layer, but they also produce hysteretic losses of their own.

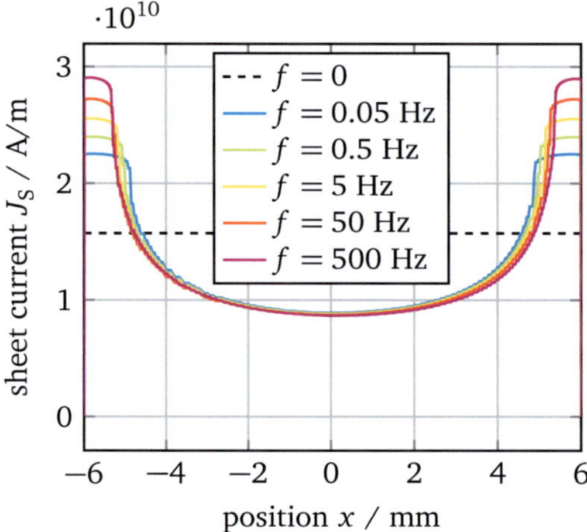

Figure 5.7: Plot of the sheet current as a function of the tape position for various frequencies in a 12 mm wide and 1 μm thick coated conductor with a critical current of 300 A and an n_p-value of 35. At high frequencies, the applied transport current flows preferably at the lateral edges of the conductor. Towards low frequencies, the current rearranges more evenly over the width of the conductor. At extremely low frequencies the current will eventually be uniformly distributed due to flux creep.

5.2.1 Influence of Permeability on Magnetic Field Distribution

For a single tape geometry with a ferromagnetic substrate, the magnetic field profiles resulting from using substrates of varying but constant relative permeabilities with analytical calculations were compared. The permeability of the substrate was varied between unity and infinity. Because the analytical solutions are incapable of calculating the losses in the ferromagnetic material, we carry this simplification over. This means that the numerical simulations also do not take the losses in the ferromagnetic material into account. We also demonstrate intermittent μ_r values to show the evolution of the field profile (see Fig. 5.8).

It is obvious that the presence of the ferromagnetic substrate slows the penetration of flux into the superconductor. This change in the magnetic field profile is a clear indicator that the ferromagnetic substrate, by influencing the magnetic field around the superconductor, also influences the superconductor itself. In reality, this interaction is mutual and complex, since both the superconductor's as well as the ferromagnet's physical properties rely on the amplitude of the local magnetic field since both the critical current as well as the permeability are functions of the field.

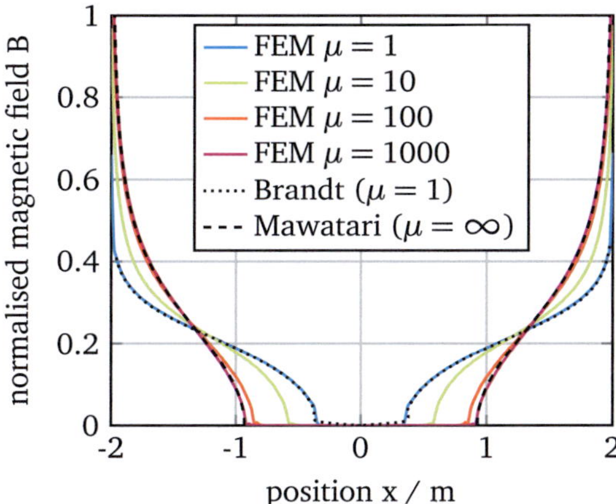

Figure 5.8: Influence of the substrate's relative magnetic permeability μ_r on the magnetic field profile with applied background magnetic fields of twice the critical field. The simulation consisted of a superconducting layer 4 mm wide, $1\,\mu$m thick with a critical current of 100 A, an n_p-value of 101 and of a substrate layer of varying but not field-dependent permeability. The substrate is $80\,\mu$m thick. Note that the field strength has been normalised as the critical tape current is of no importance in this plot. In the case of a ferromagnetic substrate, the magnetic field clearly does not penetrate as far into the superconducting strip as in the case of a non-magnetic substrate.

Keeping the limitation of not considering hysteretic losses in the ferromagnetic domain, we regard the hysteretic losses as a function of applied background magnetic field. Using these simulations to further investigate the agreement of different numerical models, we compare the losses calculated by the H-model, the adaptive-resistivity model [FFG10] and the electrostatic-magnetostatic-analogue model [GRKN09]. The agreement is generally very good, reinforcing our assumption that the H-model is reliable also when considering ferromagnetic materials [KGF11].

5.2.2 Field Dependent Current Distributions

All the numerical simulations shown previously have been run with a constant critical sheet current $J_{S,c}$. However, $J_{S,c}$ is usually dependent on the local magnetic field (see Cap. 3.1.2). This effect is important when looking at coated conductors due to their high aspect ratio (see Fig. 5.10).

As a matter of curiosity, the response of the coated conductor to applied electric transport current and applied background magnetic field loads when using the $J_{S,c}(|\vec{B}|)$ dependence (using the enhanced Kim-model, see Eq. 3.3 with parameters $k = 0.1$ and

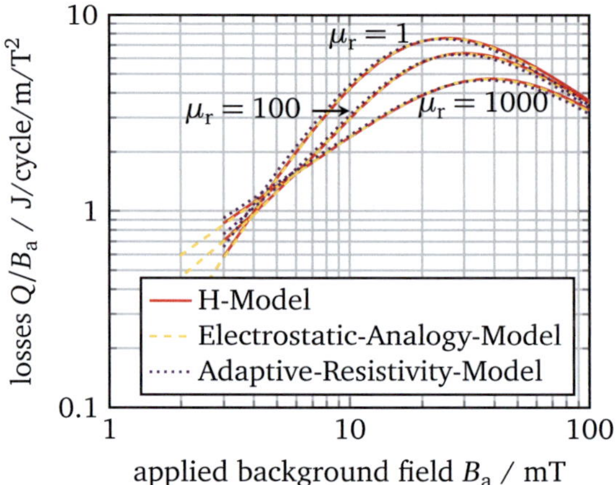

Figure 5.9: The agreement when calculating losses using simulations of various numerical models, namely the H-, the adaptive-resistivity and the electrostatic-magnetostatic analogue model, is striking, strongly hinting at the validity of these numerical models. The simulated geometry has the same properties as the one shown in Fig. 5.8. Refer to the accompanying text for tape parameters.

$b = 0.5$) is closer to the analytic solution than if constant $J_{S,c}$ were assumed. All things considered, this effect seems to be purely coincidental and may be attributed to the fact that n_p values as high as 35 allow a electric current density j higher than critical current density j_c. The field dependence on the other hand reduces the effective j_c and the resulting distribution looks similar to the analytical solution.

The differences are towards the extremal points in the graph: close to the central points where the local magnetic field vanishes (also compare Fig. 5.2, Fig. 5.3 and Fig. 5.4) the sheet current J_S exceeds $J_{S,c}$. Towards the lateral edges where the field lines are concentrated due to the high aspect ratio of the coated conductor the local magnetic field amplitude is higher. This leads to reduced critical sheet current $J_{S,c}$, explaining the reduced sheet current flowing in these regions.

5.2.3 Magnetisation Losses of a Superconductor-Ferromagnet Bilayer

The comparison of the different numerical models in Fig. 5.9 does not yet account for the losses in the ferromagnetic material itself. The electrostatic-magnetostatic analogue model in its present state is incapable of simulating the losses in the ferromagnetic material itself and is limited to modelling the changes in the electric and, by

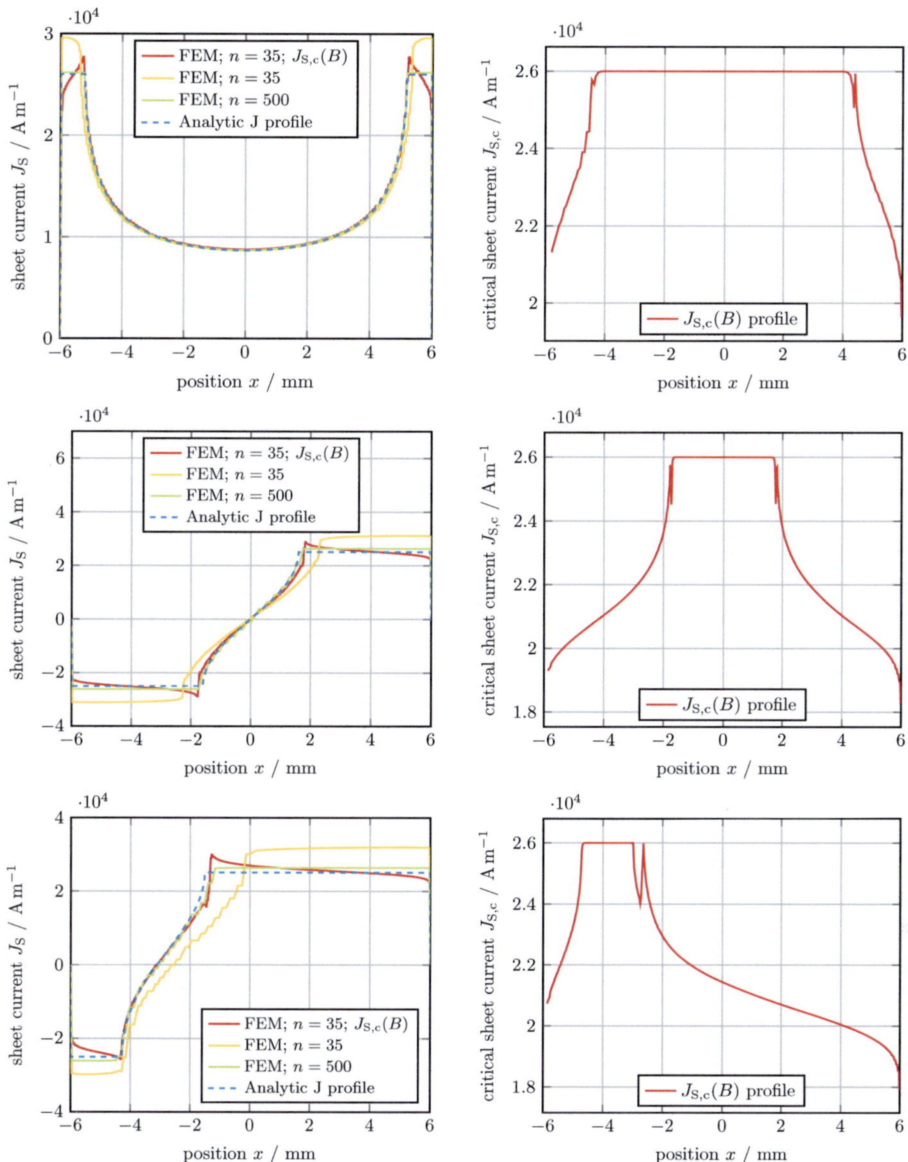

Figure 5.10: Shown to the right is the $J_{S,c}$ distribution of a field dependent critical current. To the left is the sheet current in a coated conductor for the case of applied electric transport current (top), applied background magnetic field (middle) and both (bottom). The red curves to the left are the sheet current profiles resulting from the critical sheet current distributions on the right.

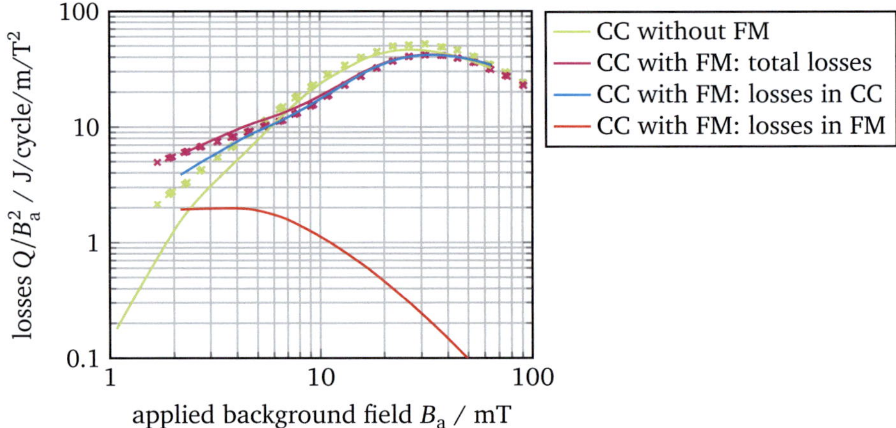

Figure 5.11: Plotted are the measured and simulated hysteretic losses in a bilayer structure consisting of a superconductor and a ferromagnet versus the applied field. The superconductor is 1 cm wide and 2.3 μm thick and has a critical current of 269.1 A. The ferromagnetic substrate is 100 μm thick. The contributions from the superconductor and those of the ferromagnet may be regarded individually. The ferromagnetic substrate increases the total losses of the assembly at low fields and decreases them towards higher fields. The measured data points were adapted and rescaled from Ref. [SIS$^+$08].

inference, the magnetic field. The goal of the comparison in Fig. 5.9 was therefore to ensure that this part of the simulation is consistent.

Ferromagnetic materials in general show a non-linear response to local magnetic fields not unlike that of superconductors (see Cap. 2.8), meaning that not only does the relative local permeability depend on the local magnetic field strength, but also the local hysteretic losses of the ferromagnet depend on both local magnetic permeability and, by induction, the local magnetic field. This is why the electrostatic-magnetostatic analogue model should only be regarded as a first order approximation and more complex approaches such as the H-model should be considered.

The physical properties of ferromagnetic materials were implemented in the H-model as outlined in Cap. 3.2.5 and compared to measured data. In the case of applied background magnetic field (see Fig. 5.11), not only are the total losses increased at low applied fields, but also the losses in the superconductor are increased by the deflection of magnetic flux by the ferromagnetic substrate. At medium to high magnetic fields, both the losses in the superconductor itself as well as the total losses are reduced when the ferromagnetic substrate is present. Note that the losses in the ferromagnetic substrate in fact do not wane but they saturate. The downward trend is solely due to the normalisation with the square of the applied background field that makes it easier to read the graph and is common in literature about hysteretic losses [GvV$^+$07, GRK11].

5.2.4 Self Field Losses of a Superconductor-Ferromagnet Bilayer

When the superconductor is supplied with an electric transport current, it is affected more detrimentally by the presence of a ferromagnetic substrate, see Fig. 5.12. Not only are the losses in the ferromagnetic substrate by itself higher than those in the superconductor with non-magnetic substrate until well above 30 %, the field shaping effect enacted by the substrate also increases the losses in the superconductor by roughly two orders of magnitude.

This effect is observable up until very high load rates. Only above 95 % of critical electric current I_c, the losses are reduced by the presence of the ferromagnetic substrate. At I_c they reach almost 20 %. This does little to remedy the detrimental effects of the ferromagnetic substrate with the exception of the unrealistic load profile with the superconductor carrying almost 100 % of its critical transport current at all times.

All presented results do little to recommend superconducting tapes deposited on ferromagnetic substrates for any of the investigated applications. Especially given the fact that the possibility to manufacture coated conductors with non-magnetic substrates exists as well. However, heterostructures of ferromagnetic materials and superconductors are not necessarily antagonistic. Arranged in carefully chosen configurations, this material combination can be very effective in enhancing the physical properties of the assembly, see Cap. 6.

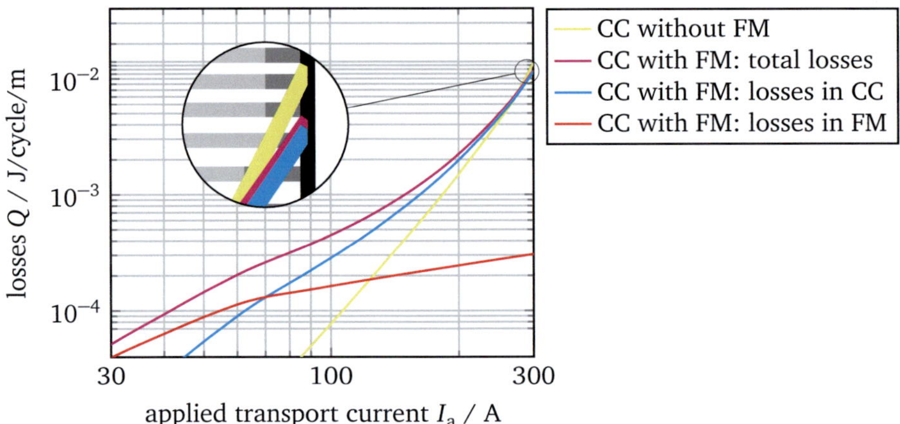

Figure 5.12: The hysteretic losses are plotted as a function of the applied field for a bilayer consisting of a superconducting and a ferromagnetic layer. The dimensions and the critical current of the superconductor are the same as in the applied field case: the superconductor is 1 cm wide, 2.3 μm thick, has a critical current of 269.1 A and an n_p-value of 35. The substrate is 100 μm thick. The presence of a ferromagnetic substrate leads to amplified hysteretic losses, both due to the rerouteing of the magnetic flux as well as the additional losses in the ferromagnetic substrate. At applied electric transport currents higher than 95 % the total losses are actually reduced, but except for this very specific and rather unrealistic case, the effect of the ferromagnetic substrate is extremely detrimental to the conductor performance.

6 Ferromagnetic Shielding

In the subsequently presented research, the influence of ferromagnetic shields on various geometric configurations is investigated. The effect of hysteretic loss reduction through the application of ferromagnetic shields for coated conductor geometries (see Cap. 6.1) as well as stacks of coated conductors (see Cap. 6.2), for bifilar coils (see Cap. 6.3) and for pancake coils (see Cap. 6.4) is demonstrated. Additionally, the hypothesis of supercritical currents is investigated for single coated conductors.

6.1 Single Coated Conductors

One of the most convenient geometries to investigate is the coated conductor, since preparing samples is straight forward as laid out in Cap. 6.1.3. Being able to compare simulation with measurement is important in order to verify the numerical model. Before coating of the samples simulations are run in order to optimise the shielding geometry to find a configuration that optimally reduces hysteretic losses. Two parameters are easily varied in the sample preparation: the coverage and the thickness of the ferromagnetic shield. For a schematic drawing of a coated conductor with ferromagnetic shields, see Fig. 6.1. A picture of actual coated conductor samples with ferromagnetic shielding is seen in Fig. 6.2.

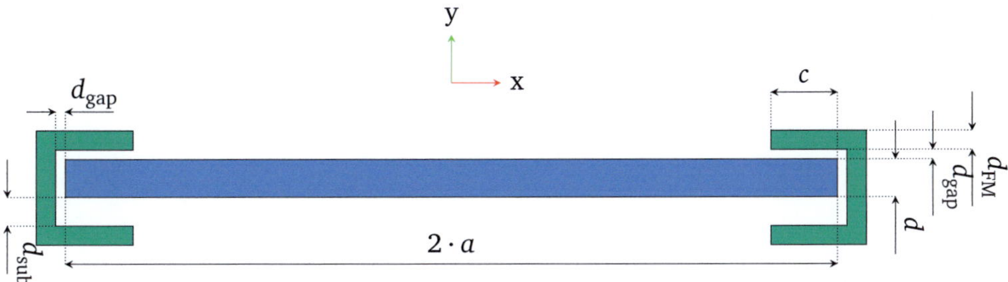

Figure 6.1: Drawing of a shielded coated conductor. The coated conductor itself has a width $2 \cdot a$, a thickness d and the shields have a thickness d_{FM}, a coverage c and they are positioned close to the superconductor with a gap d_{gap}. Instead of a substrate, there is an air gap sized d_{sub}. The substrate is usually between 50 μm and 100 μm thick but the thickness has little influence.

The coverage is the fraction of the surface starting from the lateral edges that is covered by the ferromagnetic material. The central parts are always left clear in order not to attract the magnetic flux lines towards the centre thereby speeding flux entry into the superconducting tape. This would basically constitute the detrimental effect already observed in the configuration of the bilayer structure consisting of a superconducting tape deposited on a ferromagnetic substrate. The idea of ferromagnetic shielding is to help deflect flux lines away from the superconductor thereby effectively reducing the local field around the superconductor and consequently reducing the hysteretic losses.

Figure 6.2: Coated conductor samples with applied ferromagnetic shielding. Note the uncoated areas at one end of the tapes where the electrodes were fastened onto the tapes. The coverage of the shields shown is 50 % of the total width of the coated conductors or in other words 25 % from each side.

6.1.1 Influence of Coverage and Thickness

As mentioned above, both coverage as well as thickness are easily varied; the former by simply changing the width of a plastic foil used to cover central parts of the strip during the electroplating and the latter by either increasing the current flowing in the Watts bath or by increasing deposition time (see Cap. 6.1.3). Increasing deposition time is the safer options for reasons explained in Cap. 2.8.1.

In order to reach optimal shield performance, the geometry of the shields has to be adjusted for specific load scenarios. An application operating at low load rates requires a different shield from a different application at medium or high load rates. See Cap. 6.1.2 for a more thorough discussion.

Here, we assume magnetisation load, meaning the superconductor is only subjected to a background magnetic field and no electric transport currents. In all applications in which superconductors are exposed to background magnetic fields leading to flux penetration, hysteretic losses are a major concern. Various magnetic field amplitudes are applied and the response of the shielded tape compared to the unshielded tape is considered. An optimal performing shield would reduce the hysteretic losses at all applied field amplitudes, or at least over as wide a range as possible.

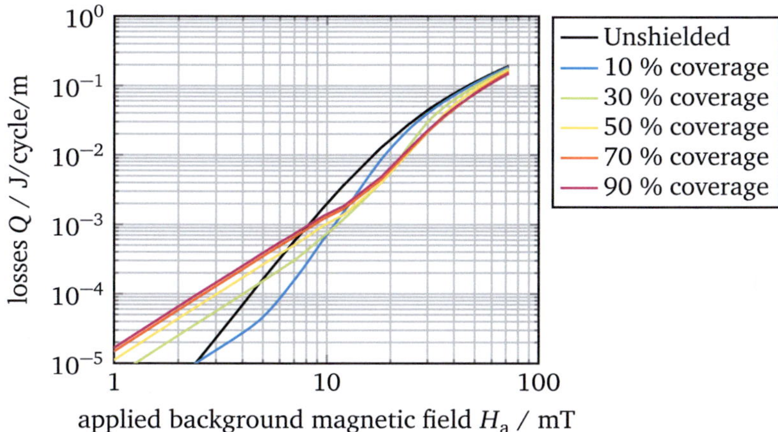

Figure 6.3: This plot shows the influence of the shield coverage for a given thickness (75 μm) of the ferromagnetic shields on the losses as a function of applied magnetic field. The superconductor is 6 mm wide, 1 μm thick and has a critical current of 87 A. The term coverage denotes the percentage of the coated conductor covered by ferromagnetic material. There is no single coverage that performs best: the ideal configuration depends on the load profile. At predominantly low loads, a small coverage is best since the reduction is greater at low fields. Towards higher loads, larger coverages result in a higher loss reduction at higher fields but also show less optimal behaviour at low fields.

Regarding Fig. 6.3 it becomes obvious that there is no single best configuration. The coverage was swept from as low values as 12.5 % to as high values as 97.5 % and there are different optimal regions for each configurations. Low coverages perform better at low applied magnetic fields but show less hysteretic loss reduction at high magnetic fields while high coverages perform better at high magnetic fields but show increased hysteretic losses at low fields when compared to lower coverages. In summary, for applied magnetic background fields, a coverage of more than 50 % does not seem advisable. For coverages between 30 % and 50 %, the reduction of hysteretic losses around 20 mT is roughly 60 %. For higher magnetic background fields around 80 mT, the savings are reduced to around 20 %.

The observed behaviour of the hysteretic losses at different coverages is due to the interaction of the hysteretic losses in the superconductor and in the ferromagnet. The hysteretic losses in the ferromagnetic shields contribute a lot at low loads, so reducing the volume of ferromagnetic material results in reduced losses at low fields. At the same time, the shielding performance is reduced at high fields because less shielding material is present. The losses in the ferromagnetic material saturate however so towards higher loads, they become small relative to the losses in the superconductor. In order to optimally size the ferromagnetic shields, the expected load profile has to be known.

In order to investigate the influence of the shield thickness, various shield thickness configurations were simulated for a number of coverages, see Fig. 6.4. Consider the first three configurations, 25 %, 50 % and 75 %: the effect of the first step to higher thickness seems to be much more drastic than the second. Even for the remaining two configurations, the effect of the second step up arguably is not as pronounced as the first. All configurations show that applying thicker shields leads to higher losses initially while at higher applied fields the losses are reduced, if only slightly.

Initially elevated total losses are in accord with expectations as hysteretic losses in superconductors are in-existent or at least negligible at very low loads. At the same time, even low loads lead to measurable losses in ferromagnetic materials. These however saturate so their influence wanes. Towards higher loads they become inconsequential when compared to the hysteretic losses stemming from the superconducting components.

6.1.2 Optimising for Various Load Profiles

General design recommendations may be given taking into account previous observations and conclusions resulting therefrom: when the expected load profile is geared towards very low fields, no shielding should be put on the conductor. At low to medium fields, the shield should be scaled up accordingly. At very high fields, an even thicker shield may prove beneficial. The same is true for the coverage: the lower it is, the more the break-even point shifts towards lower fields and vice versa. In any case, it is extremely important to know or have a good estimate of what the load profile will look like and what kind of loads are to be expected most of the time in order

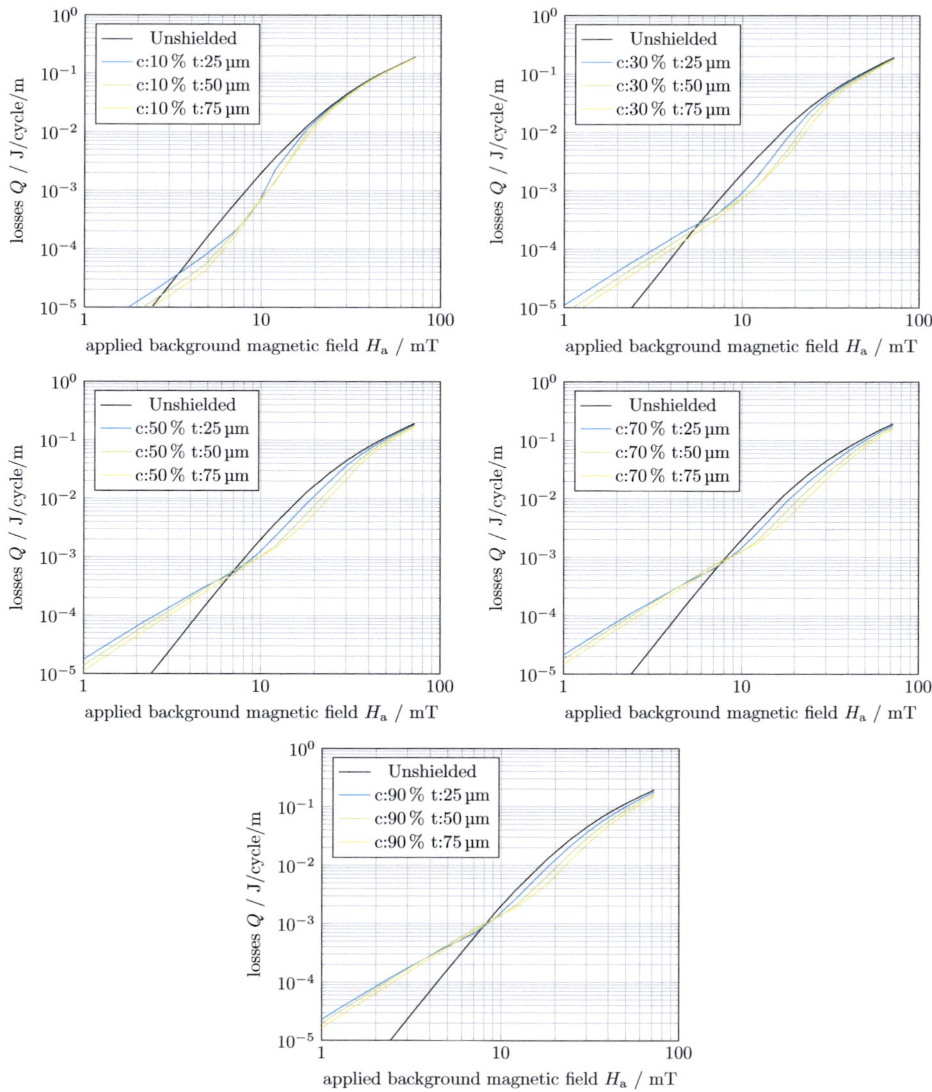

Figure 6.4: Each graph shows the influence of varying the thickness of the shields for different shield coverages on the hysteretic losses. In all cases under scrutiny, the result is generally the same: a shield of increased thickness leads to higher losses initially while at higher fields, the losses are reduced. Conductor properties are the same as in Fig. 6.3.

to optimise for that particular regime. Applied magnetic fields, electric currents and combined loads all require their own special design.

6.1.3 Electroplating Ferromagnetic Shields onto Coated Conductors

Nickel was used as the material of choice in the investigation of ferromagnetic shielding since it is both well understood, characterised and easily electroplated as mentioned previously in Cap. 2.8.1. After having established the optimal coverage and thickness of ferromagnetic shields to be measured in magnetisation measurements, samples were prepared. In order to electroplate them with Nickel, a first batch of silver stabilised Superpower *REBCO* tapes were laminated with a thin plastic foil, covering just the middle of the tape. The plastic foil was cut to correspond to the correct fraction of the tape to be left without ferromagnetic material. The plastic foil used was self-adhesive and no glue was used in its fixation to allow convenient removal of the foil at a later stage. At one end, a short length of the tape was left uncovered so the electrode required for the electroplating procedure could be connected (see Fig. 6.5).

Figure 6.5: A sample clad with copper (20 μm) instead of just being stabilised with a thin silver shunt. The blue plastic foil is applied before the electroplating procedure in order to prevent Nickel deposition in the central area.

While the sample is submerged in the bath solution and the process of electroplating Nickel is running, it is advisable to jolt the sample being coated every minute or so. This procedure helps insure a high quality deposition layer of the ferromagnetic material as little gas bubbles invariably forming on the ferromagnetic layer and clinging to the surface are shaken free. If not removed regularly, these bubbles hinder local deposition and lead to pitting in the layer and thus inhomogeneous material deposition, see Fig. 6.6.

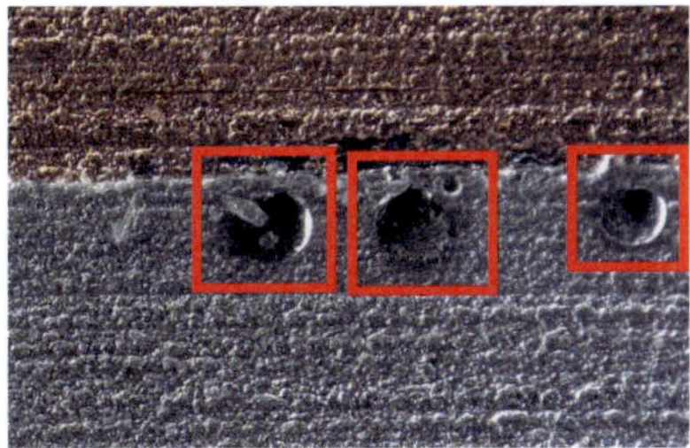

Figure 6.6: If not jolted regularly, gas bubbles cling to the freshly deposited ferromagnetic layer and cause pitting (see red boxes). The pits are about $200\,\mu$m in diameter and between $10\,\mu$m and $25\,\mu$m deep.

Hysteretic loss measurements were taken after having completed the first batch of samples and compared with numerical results from FEM modelling. Their agreement was found to be very poor which lead to extensive plausibility checks. After having verified the numerical model with test cases, every single parameter of the simulation was swept over a range of plausible values in order to find the cause of the divergence. The systematic sweep included parameters like superconductor width and height, different meshes, the response and loss functions of the ferromagnetic material, coverage and thickness and finally superconductor critical current. It turned out that with a critical current of roughly half the initial value, the hysteretic loss measurements could be matched.

The superconductor was then scrutinised for possible damage and indeed, as the plastic foil covering the central parts of the conductor was removed, portions of the silver stabilisation were found to be ablated as well, see Fig. 6.7.

Figure 6.7: A sample without copper plating shows ablating of the silver shunt. The superconductor below is also affected. Some of the damaged regions are marked red.

The acidic environment of the Watts bath is able to penetrate the porous silver layer which is only $1\,\mu$m thick and dissolves the superconductor. In order to protect the superconductor from the detrimental effects of submerging the coated conductor for some time in an acidic solution, samples clad with a $20\,\mu$m thick copper layer were used. The copper successfully hindered the solution from coming in contact with

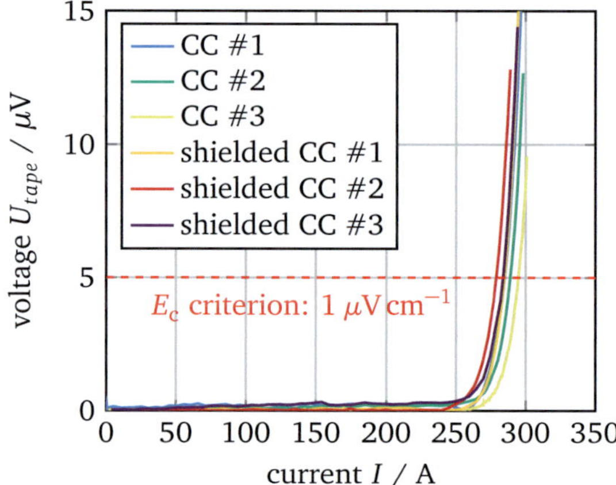

Figure 6.8: Shown are I_c measurements before and after the coating process. The samples were clad with a $20\,\mu$m layer of copper, to ensure the acidic Watts bath used in the electroplating procedure could not reach the superconducting layer. Note that no increase in critical electric current I_c is observed after coating. No considerable degradation is noticeable either, though.

the superconductor. In the new samples no degradation of the critical current was observed, see Fig. 6.8.

An increase in critical electric current I_c as predicted in Refs. [GSF00, Gen02, JJF02, JBH05, GRKN09] could not be observed in the measurements. No localised current distributions were investigated, for example using magneto-optical techniques. Since only integral currents were measured, such possibly present local supercritical currents would go unnoticed, if they do not result in a significant global increase in critical current. The measurements show however that careful handling certainly does not diminish conductor properties, either.

6.1.4 Applied Magnetic Fields

Having ascertained that the coating process itself did not diminish the critical current of the copper clad samples, we measured their hysteretic losses after the coating process at various frequencies and amplitudes. The numerical simulations were compared with the measurements taken with the combined thermal/complex susceptibility and the calibration free measuring system (see Cap. 4.2).

We simulated a Superpower 2G coated conductor that was 12 mm wide and $1\,\mu$m thick, clad with $20\,\mu$m copper and we used superconducting properties fitted to the data obtained from measurements of the bare conductor. The critical current of the tape was determined to be 286 A and the n_p-value to be 50. We further used the following parameters for the extended Kim model mentioned in Cap. 3.1.2: $k = 0.01$, $b = 0.2$. The ferromagnetic shield covered 3 mm from each side and is 30 μm thick.

Fig. 6.9 and Fig. 6.10 illustrate how the ferromagnetic shielding works and where it redirects the magnetic flux surrounding the superconductor. When reading the arrow plot, it is important to keep in mind that the arrows each show the direction of the magnetic field at their point of origin only. The ferromagnetic shield clearly attracts the magnetic field and in the lower part the arrows showing the horizontal flow of the magnetic field demonstrate the guiding of the magnetic field inside the ferromagnetic shields.

The hysteretic losses calculated using numerical simulation and the measured loss values were compared. The results show excellent agreement, with the exception of the low field regime, were we overestimated the losses in the ferromagnetic shield, see Fig. 6.12. This is probably due to the rather idealised layout of our numerical simulations: we assume a constant thickness over the wide face of the tape. However, it is well known that due to field elevation at sharp edges the electroplating process will deposit tear shaped coatings at the pointed lateral edges: the so called dog-boning effect. If the cross-section of the tape is considered, it will remotely resemble a dog-bone – hence the name. The dog-boning effect is visible on one of samples shown in Fig. 6.11. The density and physical properties of the Nickel might also not be homogeneous on top of the superconductor possibly having lower $J_{\mathrm{S},c}$ at the edges, leading to further complex interactions. Additionally, the critical current density towards the edges of a coated conductor is usually diminished as compared to the central parts of a tape.

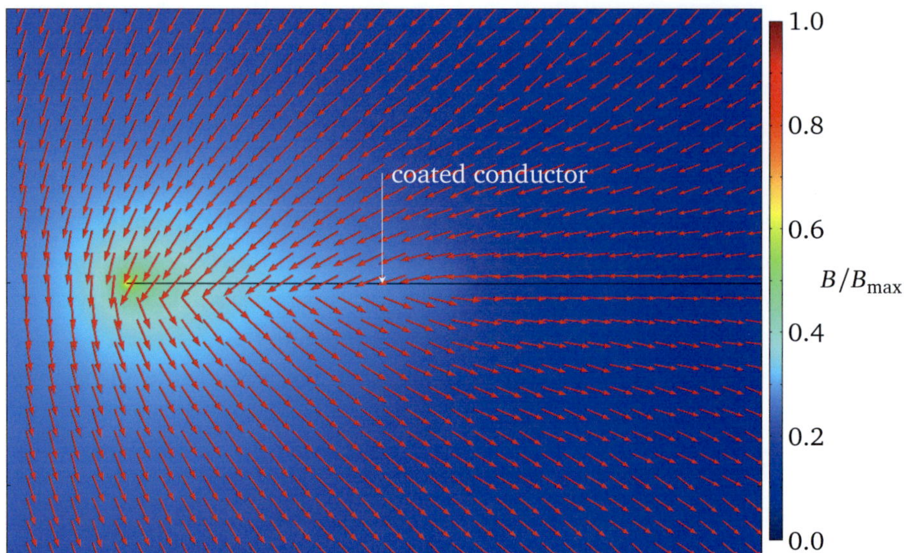

Figure 6.9: The magnetic field surrounding the left edge of a single coated conductor is shown. The arrows show the orientation of the field whereas the colour plot shows the relative magnetic field amplitude. Note the magnetic field being parallel to the superconductor where magnetic flux has not penetrated the superconductor.

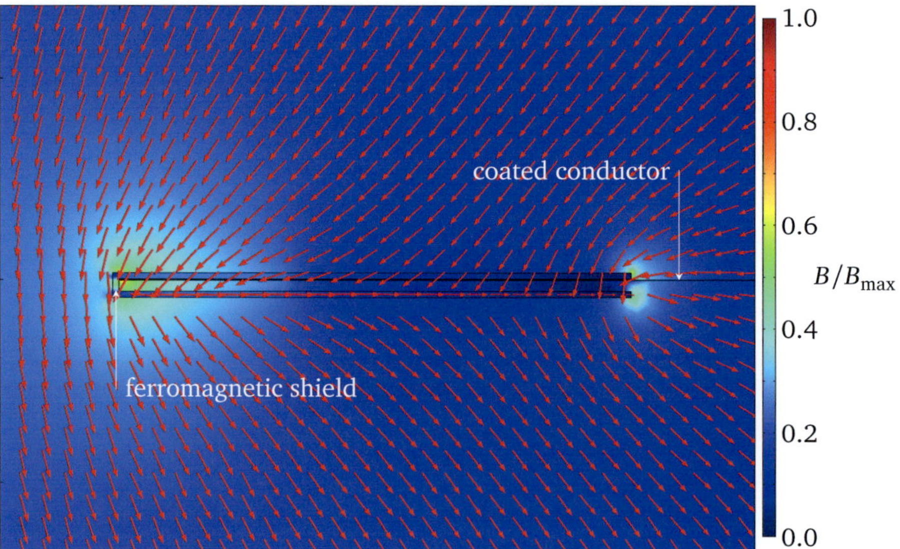

Figure 6.10: The magnetic field surrounding the left edge of a ferromagnetically shielded coated conductor is shown. The arrows show the orientation of the field whereas the colour plot shows the relative magnetic field amplitude. Note the flux rerouting in the lower part of the shield and how the ferromagnetic domain attracts magnetic flux.

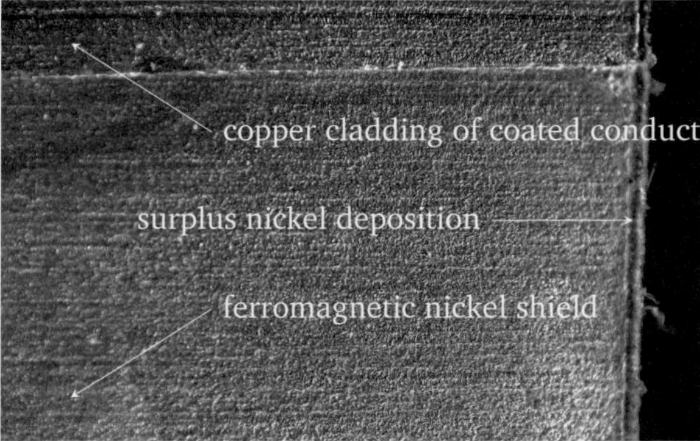

Figure 6.11: Picture of a copper clad coated conductor with applied ferromagnetic shielding. The copper cladding is visible at the top, the silver colored nickel shows a dog-boning effect at the right side.

For the applied background magnetic field simulations and experiments, the observed loss reduction was almost 50 % at 10 mm and still about 30 % at a magnetic field strength of 20 %. At higher magnetic background fields, the reduction steadily decreased. Apart from the low field region, no increase in total hysteretic losses was observed so that our data suggests always using magnetic shielding if background magnetic fields of not too high a magnitude are present.

The measurements of the unshielded and shielded coated conductor samples were taken with the calibration free measuring system. The measurements were taken with various samples and at various frequencies. Since the latter did not influence the losses per cycle, the data is shown indiscriminately.

It would be highly interesting to use materials with improved magnetic permeabilities, low loss functions and low saturation field densities. When referring to Eq. 2.15 it is obvious that Nickel's initial relative magnetic permeability of about $\mu_r = 120$ which is low to begin with decreases swiftly towards higher magnetic fields. It is decreased by more than 50 % at fields as low as 10 mT. Other materials should be much better suited to shield the superconducting volume from magnetic flux entry.

In want of a complete dataset for Ferroxcube A, we used Nickel's loss function adjusting the loss per cycle as well as the saturation flux density and used a constant permeability of $\mu_r = 1400$ with the boundary condition of applied oscillating background fields while keeping the geometric layout constant. For the loss per cycle and the saturation flux density, we used the values provided in Tab. 2.3. Exemplary simulations with Ferroxcube A as the material for the ferromagnetic shields demonstrate the high potential of using high permeability, low loss ferromagnetic materials

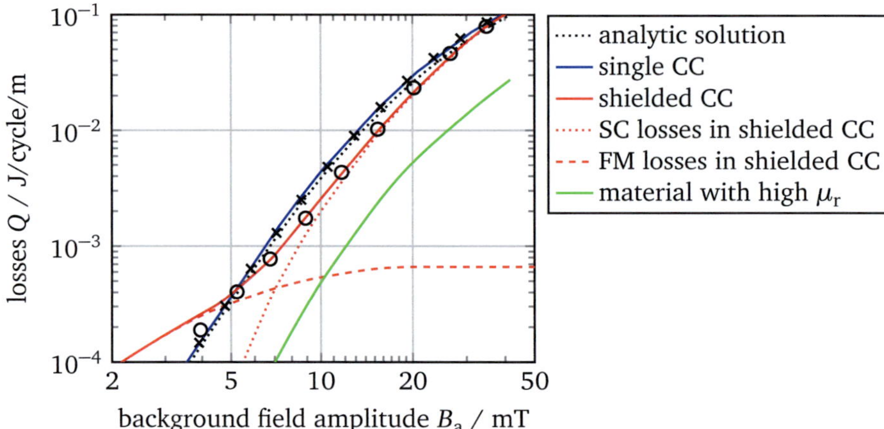

Figure 6.12: Measurements and simulations of hysteretic losses due to the application of oscillating magnetic fields; plotted are the Brandt [BI93] solution for a single coated conductor, the numerical simulations and the measurements for both the unshielded as well as the shielded coated conductor. The black crosses and circles mark measured data for the unshielded and shielded coated conductor, respectively. It was acquired using the calibration free measuring system, see Cap. 4.2.2. Using Ferroxcube A (solid green) instead of Nickel, the total hysteretic losses should be much lower due to its higher permeability (in these simulations, we used a constant permeability of $\mu_\mathrm{r} = 1400$) and lower loss per cycle ($40\,\mathrm{J\,m^{-3}}$ instead of $2.75\,\mathrm{MJ\,m^{-3}}$).

The saturation level of the hysteretic ferromagnetic losses depends foremost on the saturation field density and the hysteresis loss per cycle. There is always a trade-off when considering materials with low saturation field densities on the one hand and materials with high saturation field densities on the other: if the saturation field density is low, the hysteretic losses in the ferromagnetic material saturate faster but the shielding is not effective anymore because the relative permeability becomes unity. On the other hand, if the saturation field density is high, the shielding is effective until higher fields but the hysteretic losses in the ferromagnetic contribute more. An ideal material has a high relative permeability and low hysteretic losses per cycle. Then, even if the saturation field density is high, the hysteretic losses in the ferromagnetic material will still be low while on the hysteretic losses in the superconductor are still effectively reduced. Materials with magnetic properties better suited to shielding applications than Nickel are found in Tab. 2.3. This does not imply easy mechanical workability which should be further investigated.

6.1.5 Applied Electric Transport Currents

The same samples as in Cap. 6.1.4 were also used in applied transport current measurements. The simulations fit measured data equally well under these boundary conditions (see Fig. 6.13). Above roughly 170 A, the hysteretic losses in the heterostructure are reduced as compared to those in the pristine superconductor.

Much as is the case for applied background fields above 5 mT, an effective reduction of hysteretic loss in the heterostructure assembly is only observed above certain fill rates as the reductions in hysteretic loss in the superconductor are counteracted by additional losses in the ferromagnetic material. With the shielding geometry in use, a hysteretic loss reduction which steadily increases is observed above a threshold of 56 % of I_c. The loss reduction is around 12 % initially, in the investigated geometry at about 66 % of I_c and around 20 % at I_c. The measurements were taken with both the lock-in method as well as using high-speed digital acquisition units, see Cap. 4.3.1 and Cap. 4.3.2.

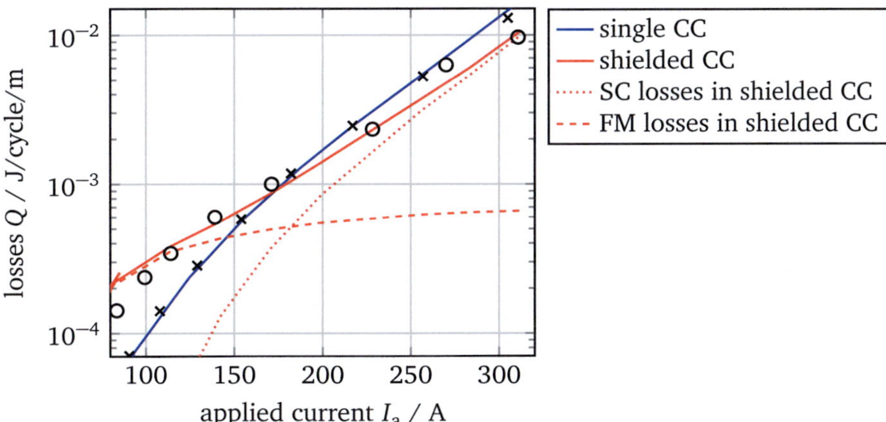

Figure 6.13: Hysteretic loss plotted versus applied electric transport current for both the unshielded as well as the shielded conductor. Above roughly 170 A the losses in the shielded coated conductor are lower than in the unshielded sample. The measurements were taken at 24 Hz for the unshielded samples and at 1.2 Hz and at 72 Hz for the shielded sample. The black crosses and circles mark measurement data for the unshielded and shielded case, respectively. The data was obtained using both lock-in as well as high-speed DAQ measurements. We did not observe any frequency dependence, ruling out eddy currents as cause for the losses.

It is worth pointing out that the total hysteretic losses at low to medium fill rates are dominated by the losses in the ferromagnetic material. By modifying the coverage and thickness of the ferromagnetic coating or using different ferromagnetic material, the shield performance can not only be tuned for specific load scenarios but also improved in general.

The samples having been specifically optimised for reduction of the hysteretic losses in background field magnetisation, the reduction in the applied electric transport current case is not optimal. However, the measurements were only used to validate the numerical model. Having established its validity by showing it correctly predicted the physical behaviour of the heterostructure consisting of a superconductor and a ferromagnet, we encourage its use to optimise geometries for specific usage scenarios.

6.2 Coated Conductor Stacks

The next logical step up from single conductors is considering stacks of tapes, since the geometry is very similar but includes more than one superconducting domain. A stack of five coated conductors was investigated with applied electric transport current and applied background magnetic field. Stacks of coated conductors by itself have an intrinsic shielding effect because the tapes at outer positions shield the inner ones. This effect raised a lot of scientific interest [GAS06, Sch06, BGN$^+$09, M99, Maw96, PS11]. For a drawing of the shielded coated conductor geometry, see Fig. 6.14. For the significance of the advanced shielding geometry, see Cap. 6.2.3.

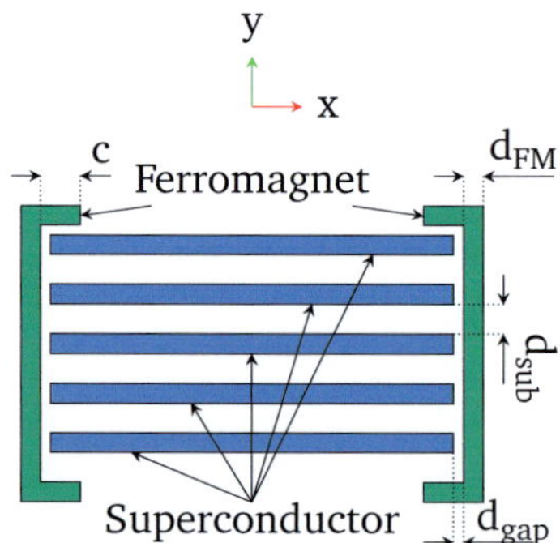

Figure 6.14: Drawing of a simple shielding geometry for a CC stack; the green parts indicate ferromagnetic domains. Their coverage c of the superconducting stack (colored blue) is varied. The thickness d_{FM} of the ferromagnetic shields is altered as well. A tiny gap d_{gap} exists between superconductors and ferromagnetic shields which is in the order of $10\,\mu$m. The spacing between two adjacent superconductors due to the presence of the substrate is d_{sub}.

6.2.1 Shielded Stack with Applied Field

A coated conductor stack of five coupled coated conductor tapes was subjected to background magnetic fields. The tapes are 1 cm wide and 1 μm thick, have a critical current density of 300 A and an n_{p}-value of 35 and are only separated by a 50 μm thick substrate.

The results of the hysteretic loss investigation show an increase in total hysteretic losses at low loads, see Fig. 6.15. With a shielding coverage of 50 % and a shield thickness of 30 μm, this translates to background magnetic fields up to 20 mT. Above, the total hysteretic losses are reduced and at 30 mT, the losses are reduced by 15 %. At 50 mT, the reduction of total hysteretic losses reaches 20 %. The hysteretic losses in the shielded sample are considerably lower than those in the unshielded sample. The reduction of the total hysteretic losses persists at about 20 %.

Up until medium fields, the losses in the unshielded sample are lower, mostly because the penetration of magnetic flux into the central coated conductors is small to begin with and the added hysteretic losses in the ferromagnetic shields cannot counterbalance the savings achieved at such low magnetic fields. Above the threshold value however where the cross-over takes place, a net reduction of hysteretic losses is achieved by ferromagnetic shielding. The behaviour has to be tuned to the load profile. Like in the single coated conductor case, lower coverage and thickness results in lower total hysteretic losses at low loads but reduced effectivity at high loads.

6.2.2 Shielded Stack with Applied Current

An electric transport current was applied to a stack of five coated conductors, treating the stack of tapes as parallel and coupled as if they were connected to the same current lead. Alternatively, this geometry is also a two-dimensional representation of a coil. Simulation of such coils is also possible, including out-of-phase excitatory electric currents.

Fig. 6.16 shows the hysteretic losses in the coated conductor stack plotted versus the applied electric transport current. The dimensions of both the unshielded as well as the shielded coated conductor stack are the same as those mentioned in the previous subchapter about applied background magnetic fields.

Above roughly 120 A which corresponds to 40 % load, the losses are reduced. While the reduction does not seem substantial in the logarithmic plot, the hysteretic loss

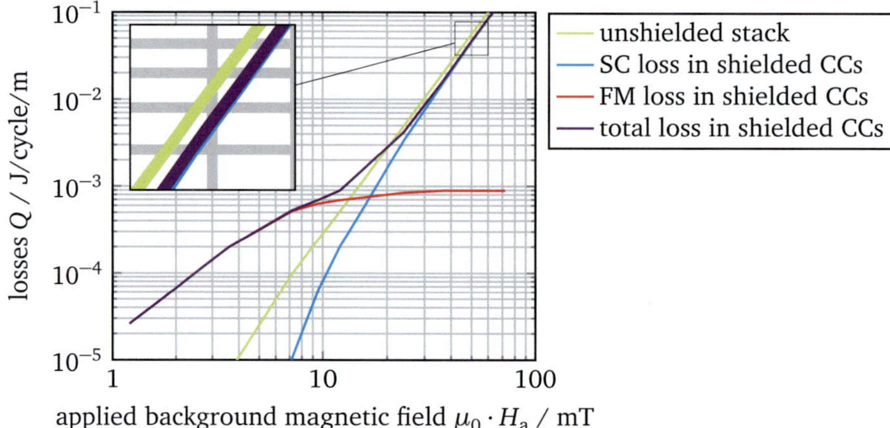

Figure 6.15: Plot of hysteretic losses versus applied magnetic background fields for both the unshielded as well as the shielded coated conductor stack. Above 20 mT the losses in the shielded coated conductor are lower than in the unshielded sample.

decrease holds steady at about 20 % as compared to the unshielded case. Again, if the level of the ferromagnetic losses could be reduced, ferromagnetic shielding would become even more attractive for a wider range of applications as the usage would be more universally applicable.

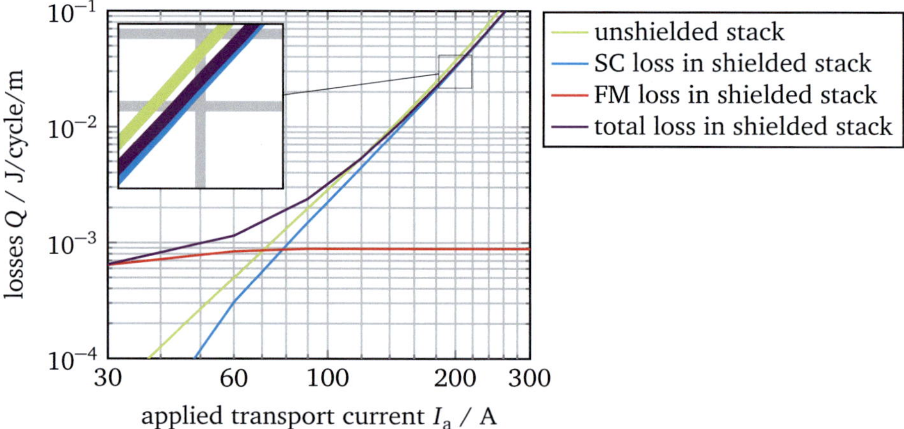

Figure 6.16: Hysteretic loss plotted versus applied electric transport current for both the unshielded as well as the shielded coated conductor stack. Above roughly 120 A the losses in the shielded coated conductor are lower than in the unshielded sample.

6.2.3 Alternate Shielding Geometry

In the first iteration, the same basic shielding layout as in the single coated conductor configuration was used, see Fig. 6.14. This is not optimal because each geometric configuration and each load case requires its own specifically optimised shielding.

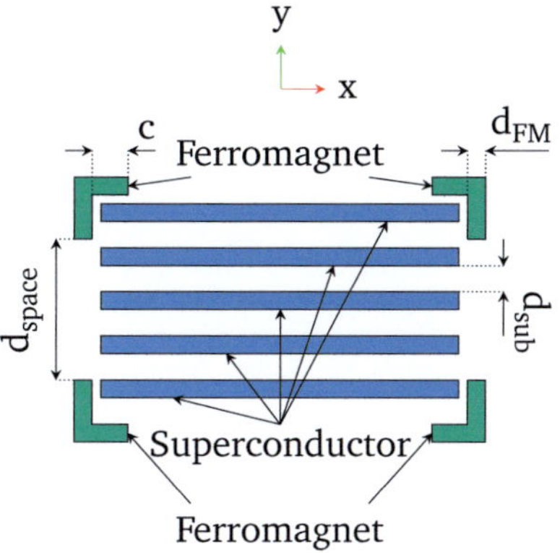

Figure 6.17: Drawing of an altered shielding geometry for a coated conductor stack. Note the missing parts of the ferromagnetic shields at the side. In addition to the dimensions mentioned in Fig. 6.14, the free spacing where the ferromagnetic shielding has been removed is given as d_{space}. In general, the shielding of a stack of coated conductors requires less ferromagnetic shield coverage than that of a single coated conductor. Removing even more of the ferromagnetic material results in a better shield performance at low to medium loads due to reduced losses in the ferromagnetic parts.

A stack of coupled coated conductors generates a magnetic field that resembles that of a bulk superconductor of the same dimensions as the stack as a whole rather closely. This means that the geometry of the ferromagnetic shields has to be adapted accordingly. A bulk has a less extreme aspect ratio which results in reduced hysteretic losses because the magnetic flux penetration is reduced.

A novel four part shielding geometry was constructed that mainly shields the corners of the stack, see Fig. 6.17. The sides of the stack are left without shields in order to reduce hysteretic losses. Since the flux does not need to be guided in those regions and the load on the superconducting tapes in the centre is low to begin with, removing the shielding there has little detrimental effect. Since the flux penetration is not as strong as in a single conductor, the coverage was reduced.

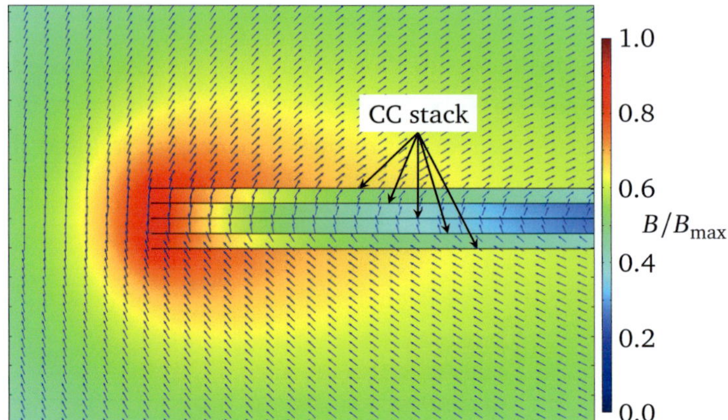

Figure 6.18: Arrow plot of the magnetic field surrounding an unshielded coated conductor stack. The colour plot in the background shows the magnetic field amplitude. Note how the field is considerably lower in amplitude inside of the stack, the outermost conductors shield the inside from magnetic flux entry.

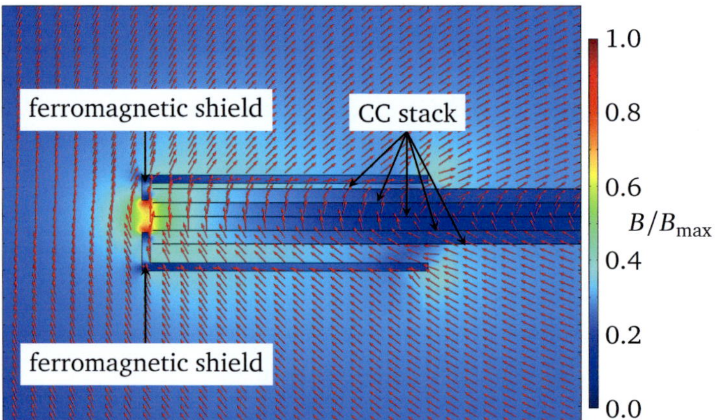

Figure 6.19: Arrow plot of the magnetic field surrounding a shielded coated conductor stack with an advanced shielding layout. The arrows show the direction of the magnetic field whereas the colour plot shows the normalised magnetic field amplitude. The optimised geometry with reduced coverage (see Fig. 6.17) is very effective in shielding the superconductors while at the same time not adding much losses on its own.

Fig. 6.18 shows the magnetic field surrounding an unshielded superconducting stack. In Fig. 6.19, the same view of a superconducting stack is shown, this time including a ferromagnetic shield of the just mentioned improved design. The strong magnetic field at the lateral edge of the superconductors where the magnetic shield is missing is clearly visible. This field would induce hysteretic losses in the ferromagnetic shielding material thereby reducing the overall effectiveness of the losses. Since the rerouteing of magnetic flux is not required in these areas, it is possible to simply remove the ferromagnetic parts there.

Looking very closely at the direction of the magnetic field shows the effect of the ferromagnetic shielding as the arrows representing the magnetic field direction are not quite as steep (the orthogonal component with respect to the superconductor is smaller) as in the case without shields. The flux rerouteing is visible inside the ferromagnetic shields where the direction of the magnetic field and hence also the arrows follow the shape of the ferromagnetic domains.

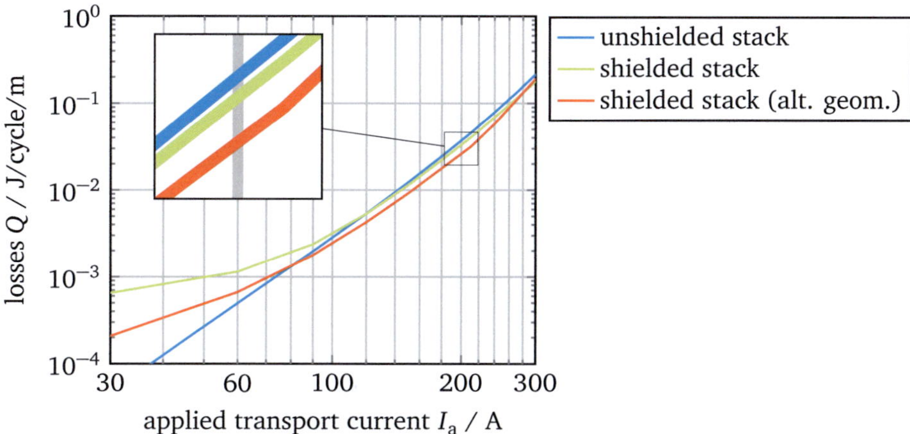

Figure 6.20: Hysteretic losses plotted against applied current in a coated conductor stack with alternative shielding geometry; the green curve shows the results of the shielding geometry shown in Fig. 6.14 whereas the red curve those of the alternative shielding geometry shown in Fig. 6.17. Note the different optimal regimes; the alternative geometry shows much earlier cross-over.

This optimised geometry performs much better at low current loads compared to the original shielding geometry, see Fig. 6.20. Only at very high electric transport current fill rates, the losses become higher than those of the regular shielded conductor stack. This is due to the central tapes carrying more current and the ferromagnetic shields not being able to shield them due to the absence of the lateral central parts. The total hysteretic losses are still lower than those of the unshielded stack, however. By adjusting coverage, thickness and the size of the gaps in the ferromagnetic shields, the behaviour can be tuned further. This again demonstrates the requirement of designing a ferromagnetic shield not only for each geometry but for a specific load case as well in order to achieve optimal performance.

In order to achieve optimal shielding performance, data on the actual load profile should be acquired. This can be used in order to apply realistic boundary conditions. A topological optimisation can then provide the best possible layout under the given constrictions. The only concern here is the speed of the simulation but with a sufficiently optimised model and starting with a rough approximation for the topological optimisation, the computational load is manageable.

6.3 Bifilar Coils

Bifilar coils are wound two-in-a-hand, meaning two conductors are used to wind the coil and are then connected in the middle. By having one conductor carry an inward and the adjacent conductor carry an outward current, their inductances cancel out resulting in a non-inductively wound coil. Another advantage is that the hysteretic losses are reduced due to the partial cancellation of the perpendicular component of the self field. Such an assembly is for example used in fault current limiters [EKF+11, EKB+12]. For a three-dimensional drawing of a bifilar coil, see Fig. 6.21.

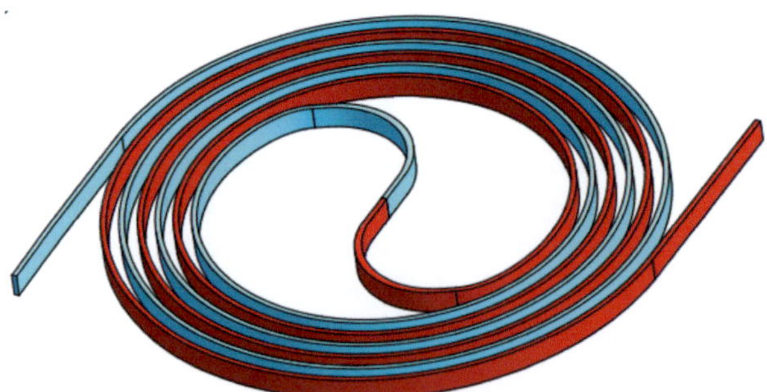

Figure 6.21: Three-dimensional drawing of a bifilar coil. The conductor changes colour from red to blue in the middle to better discern the winding. Because of the bifilar winding, each adjacent conductor carries anti-parallel current. Also see the graphic demonstrating the idealisation of the numerical model in Fig. 6.22. Picture courtesy of J. Brand, KIT.

If considered two-dimensionally and neglecting end-effects, the bifilar coil may be represented by an infinitely stacked array of superconductors and thus be modelled, due to symmetrical reasons, by a single superconductor and adequate boundary conditions. In the following simulations, a superconductor width of $2 \cdot a = 10$ mm and a superconductor thickness sample thickness d of $d = 1\ \mu$m was chosen. The separation between two adjacent superconductors was assumed to be $d_{\text{sep}} = 1.5$ mm, as reported in Ref. [EKF$^+$11].

The lines equidistant from an arbitrarily chosen superconductor with respect to both its neighbours are taken as the top and bottom boundaries of the simulation environment and the perpendicular field component is forced to zero along their course, making use of a Dirichlet boundary condition: $H_y = 0\ \forall\ y = \pm d_{\text{sep}}/2$. The Dirichlet boundary condition is used because the magnetic field component perpendicular to the tape surface generated by two adjacent superconductors carrying current in opposing directions cancels out on the demarcation line situated equidistantly from both conductors, compare Fig. 6.22.

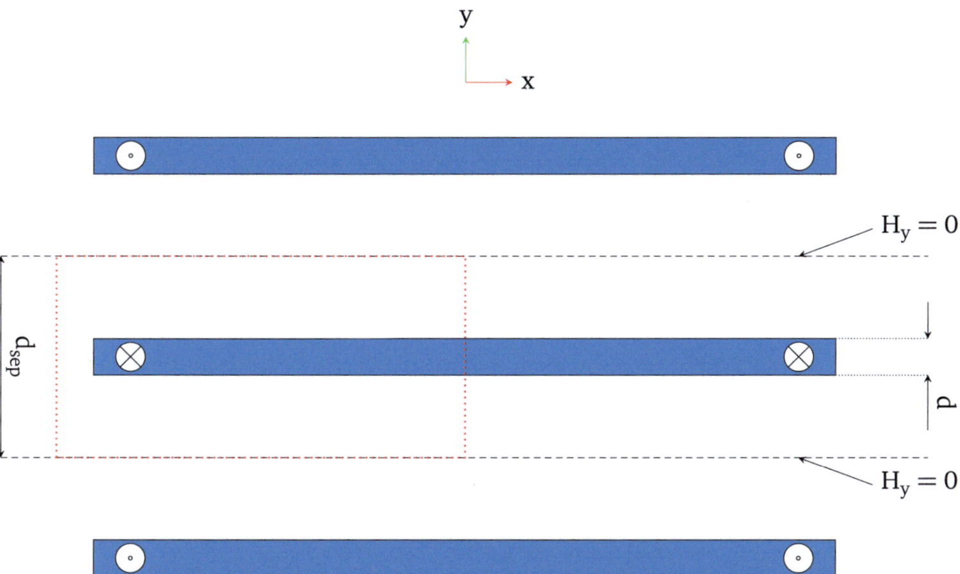

Figure 6.22: Drawing of the bifilar simulation environment with the superconductors coloured blue. The separation between two adjacent conductors is $d_{\text{sep}} = 1.5$ mm whereas the thickness of the superconductor is $d = 1\ \mu$m. The Dirichlet boundary condition along the dashed lines is $H_y = 0\ \forall\ y = \pm d_{\text{sep}}/2$. The red box sketches the region shown in Fig. 6.23.

Designing the shielding geometry for a bifilar coil requires knowledge of how the magnetic field looks like around the superconductors in order to find geometries that effectively reroute magnetic flux and lower the overall hysteretic loss. Refer to Fig. 6.23 for an arrow plot of the magnetic field surrounding an unshielded bifilar coil which is quite different from that of a single coated conductor due to the different boundary conditions (compare with Fig. 6.9).

The bifilar coil is loaded with an electric transport current. Applying a background magnetic field is nonsensical as the coil in a first approximation behaves like an infinite stack and the usage of bifilar coils is probably exclusively in applied current cases. An analytical solution for an infinite y-stack of superconducting tapes exists [Cle08]

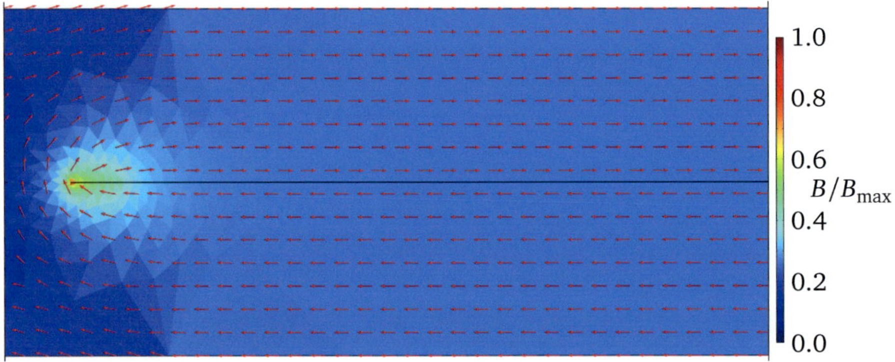

Figure 6.23: Arrow plot of the magnetic field surrounding the lefthand edge of a bifilar coil, the region shown is that encircled in red in Fig. 6.22. The arrow plot shows the direction of the magnetic field at the arrows' point of origin whereas the colour plot shows the normalised magnetic field amplitude. Note the magnetic field being parallel to the top and bottom boundary - a result of the boundary condition forcing the vertical component of the magnetic field to zero.

and may be used for comparison with the simulations in the case of absent shields, see Fig. 6.24. Note the excellent agreement apart from the deviations at regions with critical sheet current where the FEM simulation behaves differently due to a low n_p value. This behaviour is similar to that in the single coated conductor, see Fig. 5.6. The agreement is taken as proof that the Dirichlet boundary condition is enforced correctly. This further raises confidence that the results of the numerical simulations will also provide correct results once ferromagnetic shields are included.

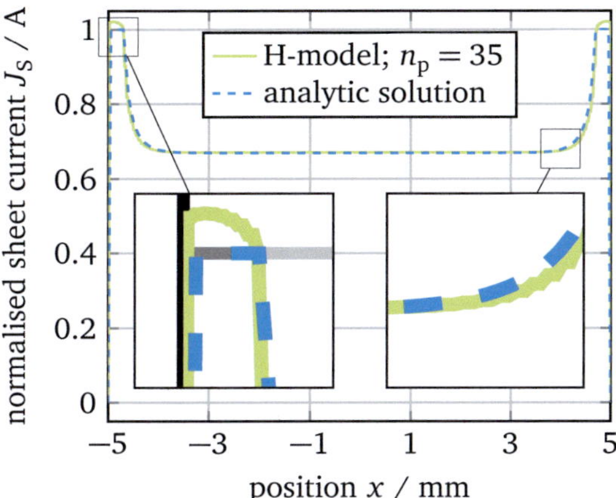

Figure 6.24: Comparison of the normalised sheet current profile resulting from the numerical FEM simulation with an analytic solution from Ref. [Cle08]. Shown are the results of a 1 cm wide and $1\,\mu$m thick coated conductor with an applied transport current of 70 % of the critical current of 300 A. The n_p-value is 35.

6.3.1 Influence of Shield Coverage

Using the same geometric layout for the ferromagnetic shielding of the bifilar coil as in the coated conductor case yields poor results, see Fig. 6.25. In order to better understand the magnetic field distribution surrounding the superconductor in the geometry layout of a bifilar coil, it is advisable to consider the arrow plot presented in Fig. 6.23.

It should be obvious that a high coverage provides little advantages since the magnetic field is parallel to the wide face of the superconductor already. Hence, a short coverage should perform equally well with regards to shaping the magnetic field while at the same time contributing much lower ferromagnetic hysteretic losses. The effect of ferromagnetic shielding is observable in Fig. 6.26.

Refer to Fig. 6.25 to see that a lower coverage indeed reduces the hysteretic losses. A lower coverage in general will lead to lower losses in the ferromagnetic parts which is why the cross-over to reducing the hysteretic losses happens at lower applied transport currents.

As is obvious from looking at the high load region, this initial reduction entails a lower shielding effect at high transport currents. Both the configuration with a coverage of $100\,\mu$m as well as the configuration with $10\,\mu$m show exceptional performance. Over

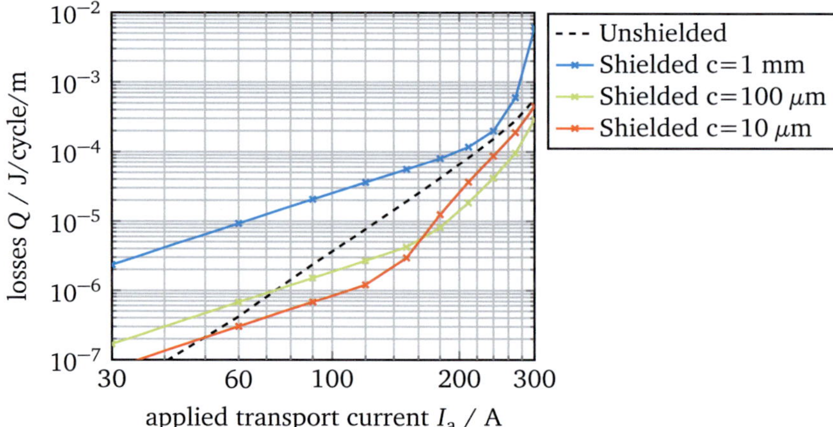

Figure 6.25: The total hysteretic loss of a bifilar coil with and without ferromagnetic shielding is compared for various coverages as a function of the applied transport current. Note the bad performance with a coverage of 1 mm on each side, corresponding to 20 % coverage. Only at considerably lower coverages the influence of the ferromagnetic shields becomes beneficial.

the whole range of applied currents the reduction is considerable but some regions even show a total hysteretic loss reduction of one order of magnitude.

A coverage of 10 μm is very difficult to obtain since the coating process is not suitable for such small coverages. Another problem are the forces occurring in real applications, especially coils at power frequency since they tend to vibrate. To mitigate these problems, slightly larger coverages can be used. Alternatively, the ferromagnetic material could just be positioned in the vicinity of the coil's turns instead of being directly applied to the edges. This should produce the same results with the added benefit of mechanical stability.

Only a small coverage is required since the bifilar arrangement itself tends to displace the electric currents towards narrow regions near the edges, as seen in Fig. 6.24.

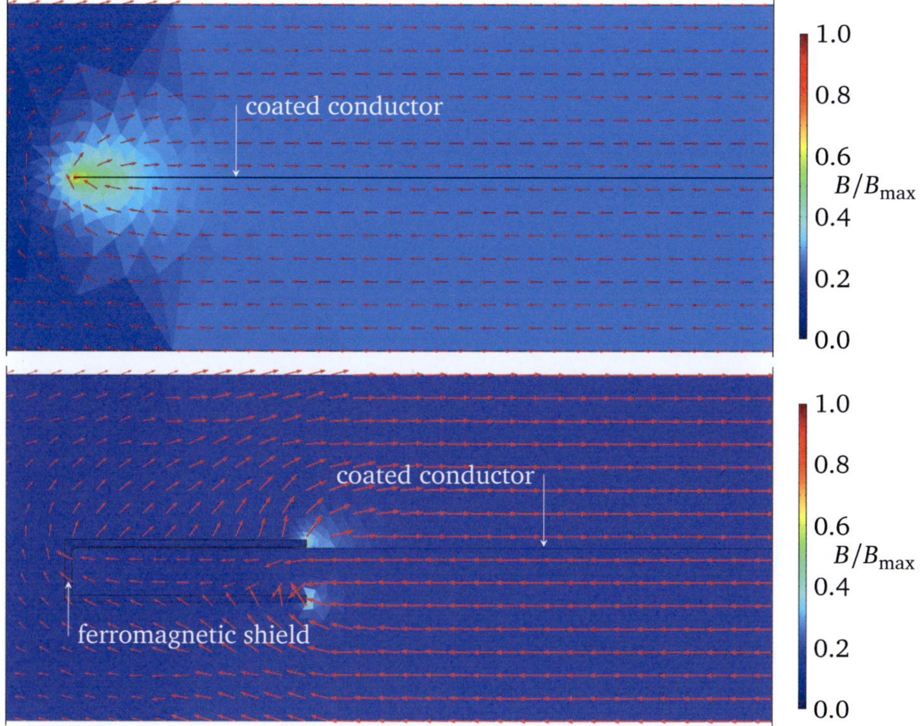

Figure 6.26: The two plots above show the magnetic field surrounding an unshielded (top) and a shielded (bottom) bifilar coil. The arrows give the direction of the magnetic field while the colour plot shows the normalised magnetic field strength. The magnetic flux is rerouted by the presence of a ferromagnetic shield. Note how the magnetic field is aligned in parallel to the superconductor anyway so the presence of the shield does little to improve the overall orientation of the magnetic field.

6.3.2 Influence of Shield Position

A further investigation targets not the coverage of the ferromagnetic shields but their placement relative to the conductor and their design. Besides testing out a ferromagnetic coating of varying thickness, the vertical position of the shields is shifted. The influence of removing the substrate is tested as well as separating the shields from the superconductor and positioning them equidistantly from both wide sides of the superconductor, see Fig. 6.27.

The total hysteretic losses of the various geometries were compared in Fig. 6.28. The configuration without a substrate is the worst at high transport current loads above 30 % I_{c} because the shields are not situated as far away from the supercon-

Figure 6.27: Drawing of a shielded bifilar simulation environment with the superconductors coloured blue and the ferromagnetic shields coloured green. The left-hand side shows a ferromagnetic shield with a substrate of thickness d_{sub}, this results in a larger gap between superconductor and ferromagnet at the bottom of the coated conductor. The separation between two adjacent superconductors (or, in this case equivalently, between the symmetry boundaries) is again d_{sep}. The right-hand side shows a layout just with gaps and without a substrate which results in a ferromagnetic shield positioned equidistantly from both wide sides of the superconductor.

ductor on one side as in the other configurations so the flux is pulled closer towards the superconductor. This configuration would be impractical anyway as a coated conductor requires some sort of carrier material and simply getting rid of it is not possible. Simply coating the edges with ferromagnetic material of varying thickness on the other hand is possible without complex processes. And these simple geometries perform very well.

Due to the special geometric configuration of the bifilar coil and the resulting boundary conditions for the numerical simulations, the hysteretic losses in these geometries are much lower than those in a single coated conductor regarded isolatedly as in Cap. 5. The shielding effect of adjacent coated conductor tapes in this geometry reduces the magnetic self field effect. Using ferromagnetic shielding only on the very edges of a tape supports this behaviour by additionally keeping the magnetic flux parallel to the wide face of the superconductor thereby delaying flux entry. This results in the large

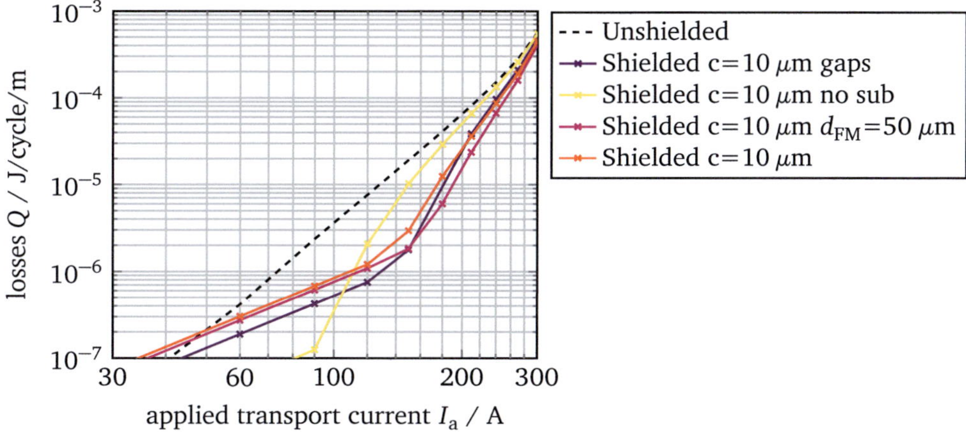

Figure 6.28: Hysteretic losses as a function of applied transport current are compared for various shield configurations. The gap configuration has a shield with equidistantly positioned parts covering the superconductor surface. Regular shields only have a tiny gap on one side where no substrate is present, see Fig. 6.27. The configuration labelled $d_\mathrm{FM} = 50\ \mu$m has thicker shields than the regular $30\ \mu$m.

reduction of hysteretic losses in the superconductor that is observed. Only relatively small volumes of ferromagnetic material are required. This leads to small additional losses which results in an effective shielding over almost the whole range of applied transport current loads.

6.3.3 Influence of Shield Thickness

Instead of varying the shield coverage or position, a configuration that proved beneficial in the previous chapters was taken and the influence of the ferromagnetic shield thickness investigated. A coverage of $10\ \mu$m with a (non-magnetic) substrate being present was chosen. After investigating the range of $10\ \mu$m to $50\ \mu$m thickness, it would seem that applying thicker shields in the bifilar coil configuration seems to always increase shielding performance, see Fig. 6.29.

It is rather interesting to see that opposed to varying the coverage, the effect of increasing the thickness of the ferromagnetic shields is always beneficial in the investigated range. Not only are the hysteretic losses reduced as the shield thickness is increased, the cross-over point is reduced. Below a certain value of applied electric transport current, the total hysteretic losses are increased because of the additional losses in the ferromagnetic material. Above this point the loss reduction in the superconductor is more substantial than the additional losses in the ferromagnetic material so the total hysteretic losses are reduced. So using thicker shields lowers the applied electric transport current cut-off value above which the addition of ferromagnetic shields is showing beneficial effects.

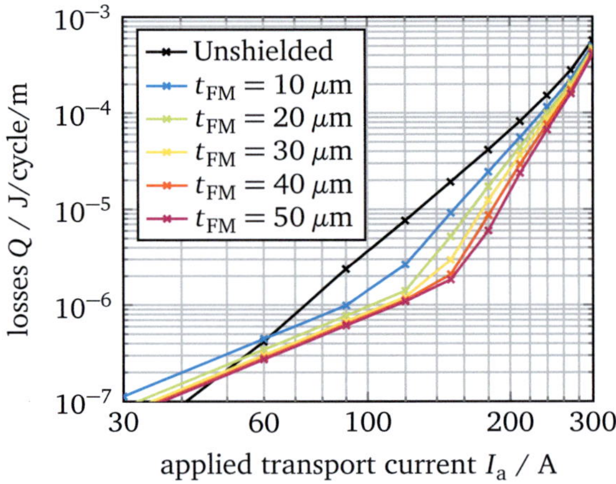

Figure 6.29: The hysteretic loss of bifilar coils with varying shield thickness is compared. The coil is wound from 1 cm wide tape with a thickness of $1\ \mu m$, a critical current of 300 A and an n_p-value of 35. The separation between two adjacent tapes is again 1.5 mm. The ferromagnetic shield coverage of the superconductor is $10\ \mu m$ in all geometries. Note the trend in the investigated range that applying thicker shields is almost always beneficial except for very low applied transport currents.

The reason of the considerable hysteretic loss reduction becomes obvious when considering the electric sheet current and the magnetic field profiles shown in Fig. 6.30 and in Fig. 6.31, respectively. The electric sheet currents being slightly lower in combination with the magnetic field component perpendicular to the superconductor surface being considerably lower in the lateral edge regions leads to considerably lower hysteretic losses. This results in the improved behaviour of the hysteretic losses observed in Fig. 6.29.

Since the ferromagnetic shielding provides beneficial hysteretic loss reduction over almost the whole range of applied transport currents, using it is generally highly recommended in bifilar coils.

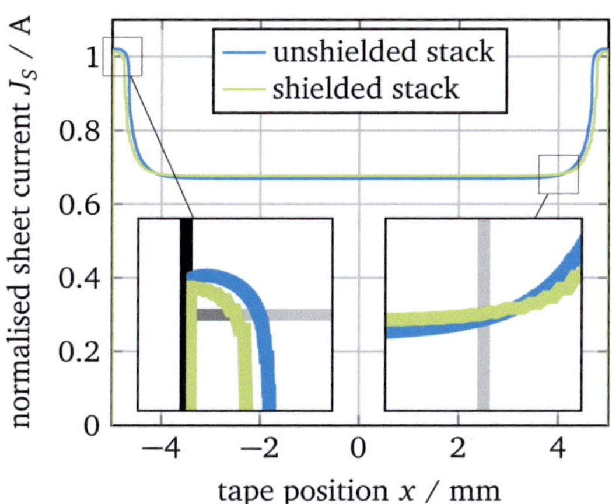

Figure 6.30: Electric sheet currents of shielded and unshielded bifilar coils plotted versus tape position are compared. Note the electric sheet currents flowing in the lateral edge regions being slightly smaller in the shielded stack geometry.

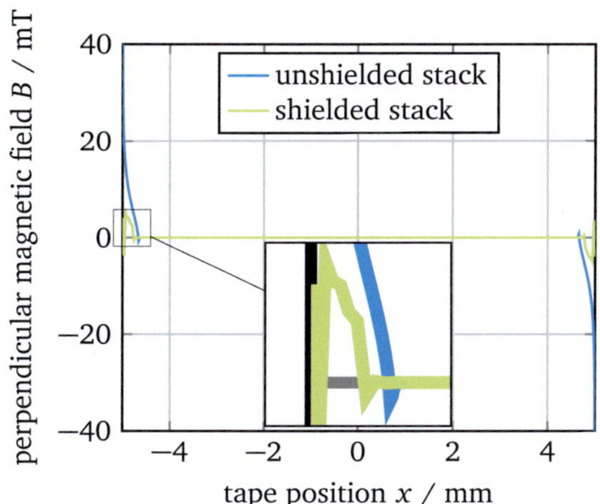

Figure 6.31: Magnetic field profiles of shielded and unshielded bifilar coils plotted versus tape position are compared. Note the perpendicular magnetic field component being much smaller in the shielded geometry. This effect in combination with the reduced electric sheet current flowing in the edge regions leads to greatly reduced hysteretic losses.

6.4 Pancake Coils

Pancake coils are closely related to bifilar coils from a geometrical point of view. The electric transport current direction in two adjacent conductors is different though: where any two neighbouring conductors in a bifilar coil carry electric currents in opposing directions, they carry codirectional electric currents in pancake coils.

This results in the boundary conditions constructed for the bifilar coil being unsuitable. Instead of the perpendicular component being zero along the previously mentioned demarcation line halfway between two adjacent conductors now the component of the magnetic field parallel to the wide superconductor face has to be forced to zero. Hence, the new Dirichlet boundary condition that needs to be enforced is $H_x = 0 \ \forall \ |y| = \pm d_{\text{sep}}/2$, see Fig. 6.32.

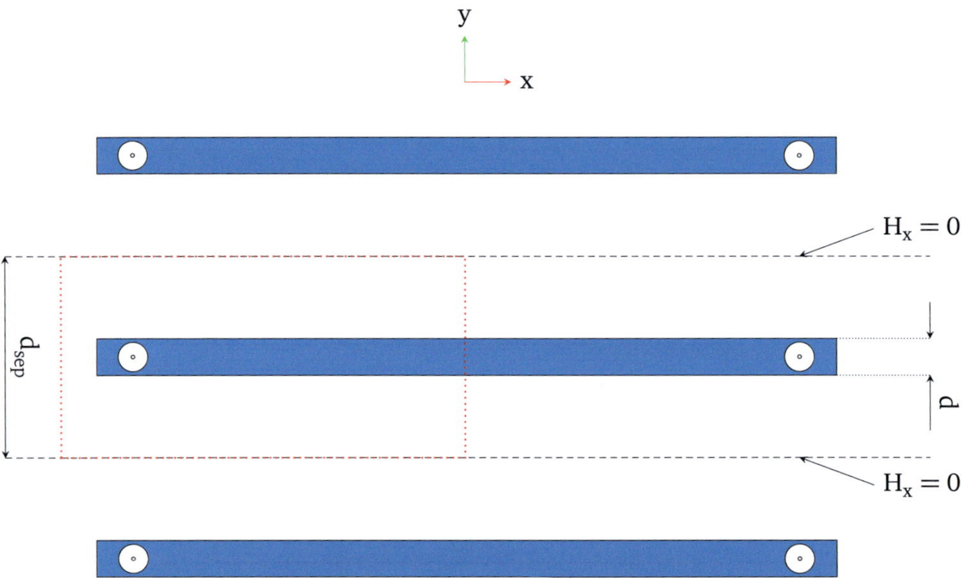

Figure 6.32: Drawing of the pancake simulation environment with the superconductors coloured blue. The geometric setup is assumed to be the same as in the case of a bifilar coil, so the separation between two adjacent conductors is $d_{\text{sep}} = 1.5$ mm and the thickness of the superconductor is $d = 1\ \mu$m. The Dirichlet boundary condition along the dashed lines is $H_x = 0 \ \forall \ |y| = \pm d_{\text{sep}}/2$. The red box sketches the region shown in Fig. 6.33.

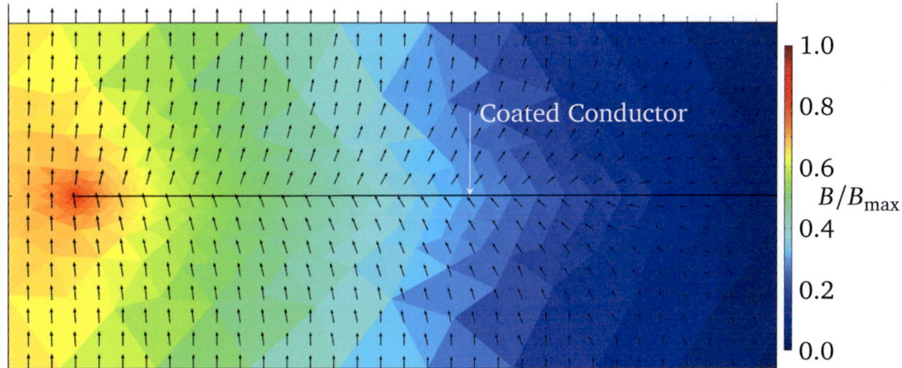

Figure 6.33: The magnetic field surrounding the lefthand side of a pancake coil, represented by an arrow plot showing direction and a colour plot showing relative magnetic field amplitude. The plotted region is marked red in Fig. 6.32. The Dirichlet boundary condition at the top and bottom forces the magnetic field component parallel to the wide superconductor face to zero.

Since the magnetic field generated in the pancake coil configuration is so different from the one surrounding a bifilar coil, it is likely the magnetic shielding has to be considerably altered.

6.4.1 Influence of Shield Coverage

As in the previous investigations, the ferromagnetic shields' coverage of the coated conductor was varied. Geometries with coverages starting from 10 μm and going up to 3 mm were tested in order to find the optimal configuration. Maximum hysteretic loss reduction was observed around 1 mm coverage which is why a detailed study was started focusing on that region, see Fig. 6.34.

The sweep demonstrates the influence of shortening or lengthening the ferromagnetic shields. The shorter the shields are, the lower are the losses at low applied electric transport currents since the ferromagnetic shields add losses themselves. At the same time, the shielding at higher load rates becomes less effective. Conversely, the longer the ferromagnetic shields are, the higher the losses due to increased losses in the ferromagnetic material and the more effective at higher load rates.

The hysteretic loss reduction due to the presence of the ferromagnetic shields in the pancake geometry is very interesting because it is manifest over almost the whole load range. Even though the reduction becomes less important towards higher loads, it is always present and at low to medium applied transport currents, it is pronounced.

From about 15 % up to 30 % I_c, the hysteretic losses are reduced by more than 50 % and at 50 % I_c, the reduction is still about 25 %.

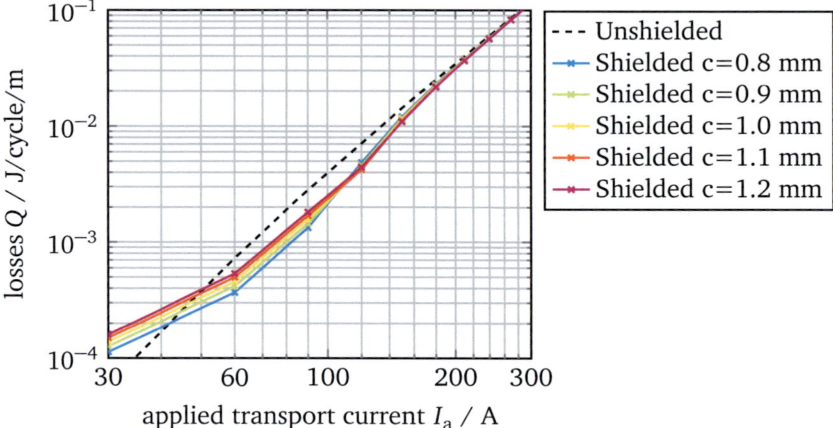

Figure 6.34: The influence of various shield coverages on the hysteretic loss behaviour of pancake coils is compared under applied electric transport current. The coated conductor is 1 cm wide and 1 μm thick, has a critical current of 300 A and an n_p-value of 35. The spacing between two adjacent tapes is 1.5 mm.

How the ferromagnetic shields reroute magnetic flux in the pancake coil geometry is demonstrated in Fig. 6.35. The magnetic field distribution demonstrates why the losses in the pancake coil geometry are so much higher than in the bifilar coil geometry: coated conductors are much more sensitive to perpendicular magnetic fields and the boundary conditions force the field to be perpendicular very close to the superconductor. In the bifilar configuration, the boundary condition forces the perpendicular component of the magnetic field to zero which reduces the perpendicular magnetic field on the superconductor surface considerably.

6.4.2 Influence of Shield Thickness

The configuration with a 1 mm coverage was selected since the hysteretic loss reduction turned out promising in the investigation of the shield coverage. For this coverage, the influence of the shield thickness was investigated. The peculiarities of the geometric boundary conditions seem to favour thick ferromagnetic shields. Even though a larger volume of ferromagnetic material implies higher hysteretic losses in ferromagnetic domains, this is apparently counterbalanced by a reduction of hysteretic

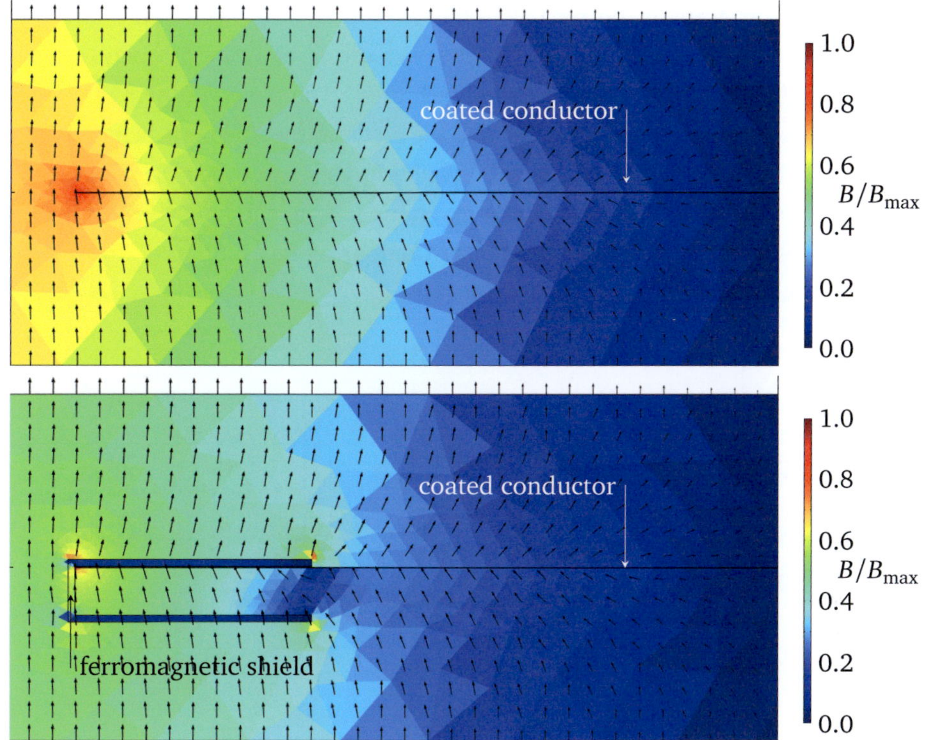

Figure 6.35: The magnetic field surrounding the lefthand side of a ferromagnetically shielded (bottom) and an unshielded (top) pancake coil is shown. The arrow plot shows direction and the colour plot shows relative magnetic field amplitude. The plot only shows a fraction of the simulation geometry, the region is marked with a red dotted rectangle in Fig. 6.32. Note how the presence of the shield changes the magnetic field distribution around the superconductor.

losses in the superconducting domain resulting in an overall total loss reduction, see Fig. 6.36.

One aspect is different from the results observed in Cap. 6.3.3: the losses at low load factors are increased as the thickness is increased. In the bifilar coil configuration, the losses were reduced at low applied electric transport currents as well. The total hysteretic losses are reduced above 45 A which corresponds to 15 % of the critical current. At as low as 60 A (20 % I_c), the ferromagnetic shielding reduces the hysteretic losses by about 40 %. At about twice the applied current, around 40 % I_c, the hysteretic losses are reduced by about 60 % as compared to the unshielded geometry. Above 66 %, the ferromagnetic shielding is less effective and only a reduction of the hysteretic losses of about 10 % is observable.

The different behaviour of the pancake coils as opposed to the bifilar coils stems from the different current flows which result in the already mentioned different boundary conditions of these two coil configurations. The observed higher losses at low loads in the pancake geometry stem from the fact that the magnetic field is aligned perpendicular to the superconductor to a much higher degree than in the case of a bifilar coil. This leads to an earlier rise in magnetic field at the lateral edges leading to higher losses in the ferromagnetic shields. Comparing Fig. 6.36 and Fig. 6.29 the difference in magnitude is immediately obvious.

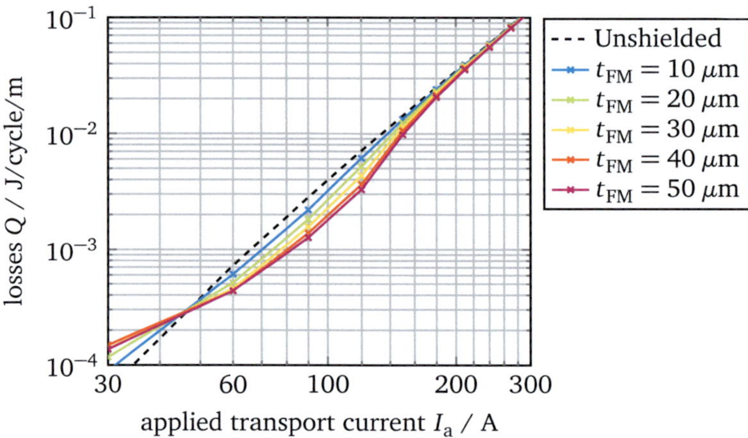

Figure 6.36: The influence of varying the shield thickness on the hysteretic loss behaviour of pancake coils is compared under applied electric transport current. The coated conductor tape is 1 cm wide and 1 μm thick, has a critical current of 300 A and an n_p-value of 35. The gap distance to the adjacent tape is 1.5 mm. Thicker shields lead to higher losses at low load rates and to lower losses at higher load rates.

6.5 High Field Coil

All superconducting coils suffer from the self field effect suppressing the critical current density on the outlying turns thereby reducing the overall current and, by inference, the magnetic field. If the self field effect could be abated, the coil would become more effective, independently of the material used. In previous chapters the potential of ferromagnetic materials to reroute magnetic flux was explored. The effect was utilised in order to reduce hysteretic losses but we propose it is also possible to increase coil performance employing the same principles.

Various shielding geometries are tested, resorting to a similar boundary condition as in the geometric configuration of the bifilar coil. This time however, a quasi infinite half-space is simulated to model the physical properties of a cylindrical coil, see Fig. 6.37.

The reason this geometric design is able to decrease hysteretic losses is the same it is able to increase coil performance: the abatement of magnetic field leads to less self field degradation and less critical current depression which in turn allows higher total currents to pass through the coil.

As with previous geometries, it is important to optimise the shielding geometry in order to improve the coil performance. We therefore investigate the influence of various shield shapes, coverages and variations in thickness in the following sections.

The coil being investigated consists of five turns of a 4 mm wide coated conductor of 1 μm thickness with a critical current of 300 A and a power-law exponent of 35. We use the advanced Kim model formula stated in Eq. 3.3 for modelling the critical current dependence on the magnetic field with parameters $b = 0.3$ and $k = 0.1$. These values are in accordance with previously published works and represent typical values for high-T_c superconducting tapes [GvV$^+$07, GvS$^+$08].

The electric transport current is applied as a 10 s linear ramp. The current in each tape is acquired with the help of an integration operator on the current component flowing along the longitudinal direction of the tape. The voltage drop per length unit is extracted utilising an averaging operator on the component of the electric field along the longitudinal direction. The simulated coil has an arbitrarily chosen bore of 2 cm. Different bore sizes were also investigated and results did not differ qualitatively.

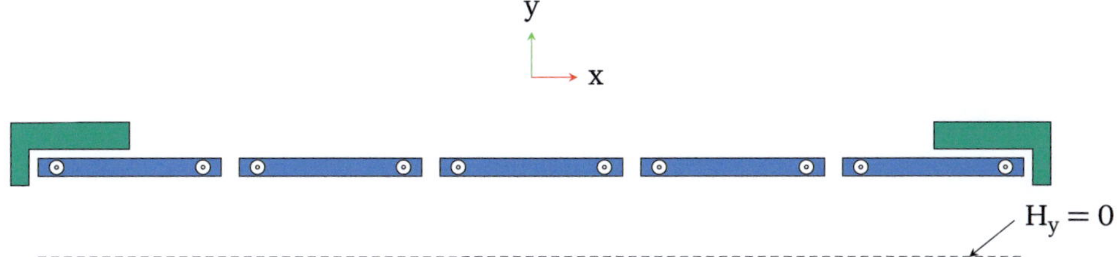

Figure 6.37: Drawing of the high magnetic field coil simulation environment with superconductors coloured blue and ferromagnetic shields coloured green. The boundary condition at the lower simulation boundary is similar to those in the bifilar coil simulations. However, only one boundary condition is required as the top half space is not constrained. The Dirichlet boundary condition along the dashed lines is $H_y = 0$.

The shield shape was investigated which was varied between L- and C-shapes of varying coverage and thickness. The C-shaped shields proved to have worse performance

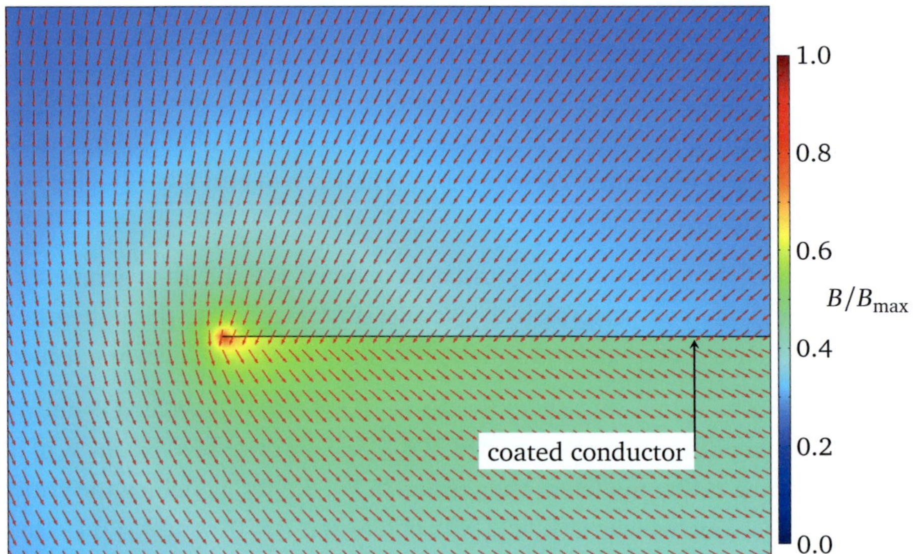

Figure 6.38: Combined arrow and colour plot showing the direction and relative amplitude of the magnetic field surrounding the outermost tape of a high magnetic field coil. Note the superelevation of the magnetic field at the lateral edge of the tape.

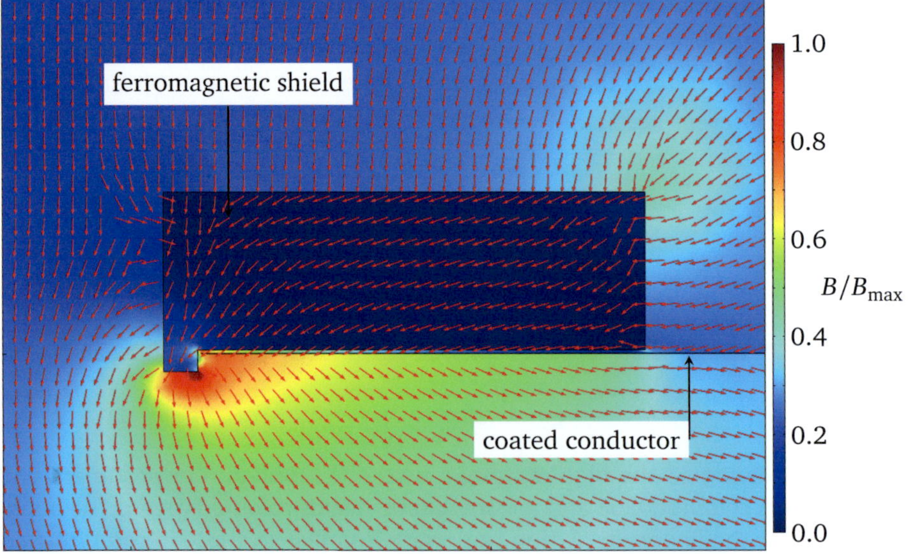

Figure 6.39: Combined arrow and colour plot showing the direction and relative amplitude of the magnetic field surrounding the outermost tape of a ferromagnetically shielded high magnetic field coil. Note the considerably reduced magnetic field component orthogonal to the superconducting tape.

than the L-shaped shields in all investigated configurations. The shield shape was therefore fixed to the L-shape with the short part of the shield extending just 50 μm into the bore. Only the two outermost tapes are covered with the ferromagnetic shields as shields also covering the tapes situated in the inner turns did not improve the coil performance. The shields are shaped as shown in Fig. 6.39 with the ideal coverage of the wide face of the superconductor being about half the width of the superconducting tape.

Having established the principal superiority of this shield design, the thickness of the shield part covering the wide superconductor face was altered from 100 μm to 900 μm. Coverages smaller or larger than the just mentioned values are neither practical nor functionally interesting for the coil under consideration. Their investigation is therefore omitted.

The FEM source code used to reproduce the simulations in Fig. 6.40 is found in Cap. A.1.2.

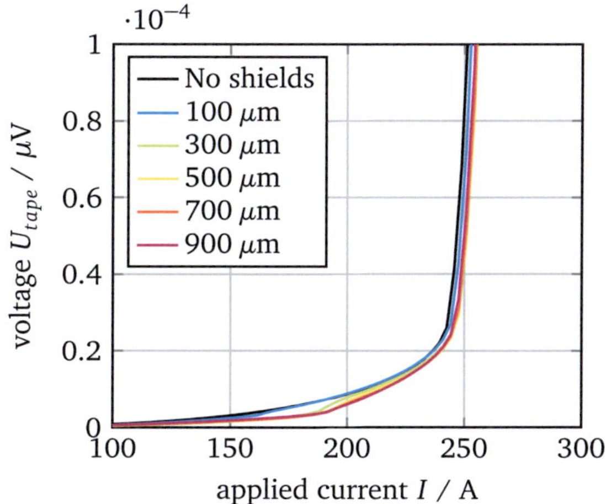

Figure 6.40: The voltage drop of an arbitrary tape length is plotted as a function of applied current. The tapes situated in the middle of the coil are hardly influenced by the presence of the shields. Therefore only the response of the outermost tape is shown. Since the self field effect is strongest for the extremal tapes, this is sufficient.

Since the maximum attainable coil performance improvement is negligible at least in the specific coil and model under consideration, this concept is at present not further investigated. The case may be entirely different for larger coils, insets and for different shielding materials. Further thorough research is therefore required before altogether abandoning the idea of ferromagnetic shielding in superconducting high field coils, especially since the field homogeneity may be favourably influenced.

6.6 Effectivity of Shielding

In Tab. 6.1 the effectivity of ferromagnetic shielding is summed up for all previously investigated geometric configurations. With the exception of the high field coil where the ferromagnetic shielding was supposed to mitigate the critical electric current suppression, the ferromagnetic shielding is a concept worth investigating.

The high field coil geometry seems not to benefit significantly from the ferromagnetic shielding. The goal when applying ferromagnetic shielding to bifilar and pancake coils, to tapes and to stacks of tapes was a reduction of hysteretic losses which was reached. In the case of the high field coil, the shielding was supposed to reduce critical electric current suppression. This appears to be almost ineffective, at least with the parameters used in the investigations in Cap. 6.5.

Whether the shielding concept is correct is impossible to say however since it was later found that the H-model is incapable of observing these supercritical states due to the reliance on the power-law. In order to investigate the concept as proposed in [GSF00, Gen02, JJF02, JBH05, GRKN09], further research is needed. The flux entry is delayed in the investigated geometries using ferromagnetic shields and the local magnetic field strength is reduced by a small amount. This leads to a slight decrease of the critical electric current suppression in superconducting materials.

Future research needs to investigate these geometries, possibly using an enhanced electrostatic-magnetostatic analogy model. In its present state, this model can only account for one superconducting domain.

Shielding effectivity		
Geometry	Shielding effectivity in the case of	
	applied current	applied field
Single Coated Conductor	good above medium loads	good at medium loads
Coated Conductor Stack	good above low to medium loads	good above low to medium loads
Bifilar Coil	very good from low to high loads	not tested, no useful application
Pancake Coil	very good from low to medium loads, low at high loads	not tested, no useful application
High Field Coil	ineffective	not tested, no useful application

Table 6.1: Effectivity of all investigated geometries with transport and field load. Only a very rough overview is given as the performance depends strongly on the load profile and the geometry.

7 Twisted Stacked Tape Conductor Cable 3D Model

The Twisted Stacked Tape Conductor cable is a high current cable with a scalable design that was developed for fusion applications [TCBM11]. It has a helical structure and utilises multiple coated conductors in order to reach the required high current densities, see Fig. 7.1 and Fig. 7.2.

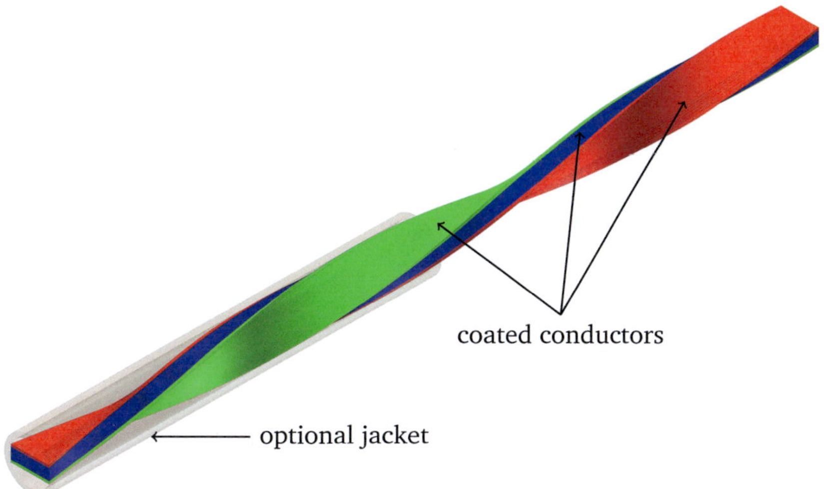

Figure 7.1: Three-dimensional drawing of a single Twisted Stacked Tape Conductor. Reproduced with kind permission from [Bar13].

A numerical model was developed to investigate the electromagnetic behaviour of the TSTC cable. The measurements of the TSTC cable were conducted using an 80 cm long sample. The DC transport current measurements are only directly comparable to numerical simulations if the latter also take into account end effects like the terminals of the cable: the contact resistances between the coated conductors and the copper contacts have to be accounted for.

The application of transport current is possible in two dimensions without losing information, assuming the cross-section is an accurate 2D description of the originally helical 3D geometry and the current distribution is not influenced by the twisting. Because of the twisting of the cable, applying a background magnetic field is not as easy. A numerical model which aims at successfully simulating the cable therefore has to be three-dimensional. Since the contact resistances influence the tape behaviour a lot especially in short samples, they need to be taken into account as well.

Figure 7.2: Picture of a section of a 2 m long TSTC cable with 32 tapes and a twist pitch of 20 cm. Image courtesy of M. Takayasu, MIT.

The complex geometry of the TSTC cable with translational symmetry, helical structure and contact resistances is therefore a perfect match for a first application of the three-dimensional implementation of the H-formulation with the newly developed boundary conditions. The numerical model is capable of simulating realistic three-dimensional superconducting assemblies and capable of including contact resistances up to date. To the best of our knowledge this combination has not yet been shown before. Note should also be taken of the computational efficiency: a simulation of a Twisted Stacked Tape Conductor cable in full three-dimensional complexity under applied electric current or background magnetic field load including contact resistances takes less than a day.

High performance superconducting magnets and coils require high-T_c superconductors due to their potential for high field, high current application. They are also interesting for running systems above liquid Helium temperature. Both first generation as well as second generation high-T_c superconducting tapes are not ideally suited for cabling due to their flat geometry. In general, round wires like in conventional conductors would be preferable. Using superconducting tapes is still possible but the cable design has to account for the flat geometry.

Apart from the TSTC cable, several other designs exist that use coated conductors. Even though the coated conductor tapes do not have an ideal geometry, their potential for high current density, high temperature applications makes them interesting. The expected low cost will be another advantage. There are several regular transmission line power cables which all carry below 5 kA, see for example Refs. [KKT$^+$99, KWN$^+$01, Mas02, MKY$^+$02, HHJ$^+$04, XHB$^+$04, KJL$^+$05, SCK$^+$06, CBK$^+$06, Mal06, DSJ$^+$07, TYM$^+$07, SHL$^+$07, MSB$^+$07, Tsu08, YAI$^+$09, MSBW09, MFL$^+$09, MKH$^+$09, SVF$^+$10, SMN$^+$12]. Of the dedicated high current cables, the most prominent are the Twisted Stacked Tape Conductor (TSTC) cable [TCBM11], see Fig. 7.1, the Conductor On Round Core (CORC) cable [vdL09, vdLLG11], see Fig. 7.3 and the Röbel cable [GNK$^+$06], see Fig. 7.4.

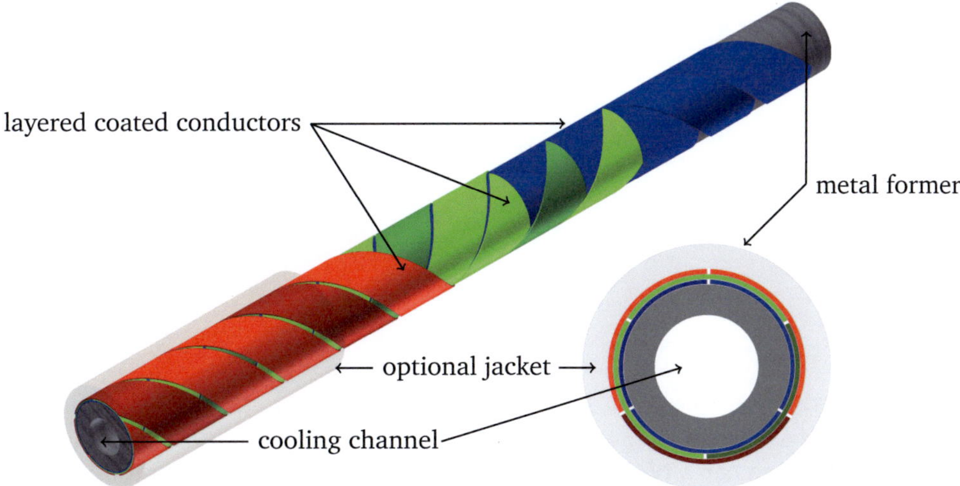

Figure 7.3: Three-dimensional drawing of the Conductor On Round Core cable. Note how the diameter of the outer layers is necessarily different from the inner ones, resulting in a different twist pitch. Reproduced with kind permission from [Bar13].

The transmission line power cables are mainly optimised for medium current, high voltage applications. The CORC cable is geometrically akin to these transmission line power cables with its coaxially wound multilayer design. It has already been used for coil construction and showed not only good electrical and magnetic, but also good mechanical stability [vdLNM$^+$13, Bar13]. In the CORC design, the coated conductor tapes are wound around a central former with an arbitrary twisting angle that depends on the number of tapes and other factors. Usually, the twist pitch in each layer is different. For fusion and AC applications, this difference in twist pitch becomes a grave problem because of the inductivity mismatch and high losses resulting therefrom. The CORC cable design is therefore preferred for DC applications.

The Röbel cable is very well suited for AC applications as all the tapes necessarily have the same twist pitch due to the transposition. Another advantage of this design is the identical behaviour of all the tapes as there are no distinguished tape positions due to the tape transposition. The Röbel cable however is mechanically more challenging as the stamping and the meanderlike structure of the tapes introduces positions of superelevated mechanical stress. Moreover, the flexibility is constricted as the cable

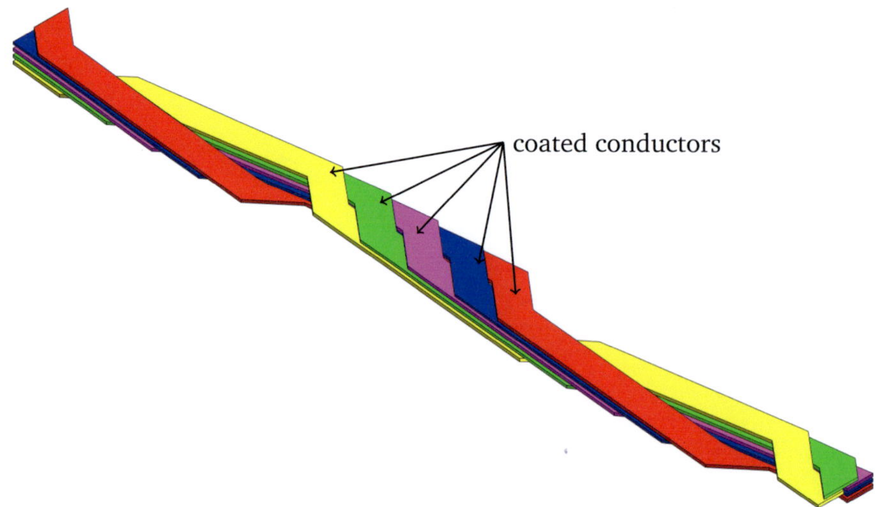

Figure 7.4: Three-dimensional drawing of the Röbel cable. Note the meanderlike structure and all tapes taking every possible position eventually. As a consequence, in cables longer than half a twist pitch, all tapes will behave identically. Reproduced with kind permission from [Bar13].

has only one almost unconfined degree of freedom for bending. The cost factor is also higher for the Röbel cable since the meanderlike tape structure entails a higher material requirement. The meanderlike structure of the Röbel cable is produced by stamping coated conductors. This leaves some unusable pieces to be discarded as waste.

The TSTC cable is mechanically flexible like the CORC cable and is not fully transposed either. It is an easily manufactured high current cable with good scale up potential. The possibility to enclose it in a round copper tube is extremely interesting for mechanical reasons.

Regarding subsequent investigations, there are two cases to be considered: (quasi-)DC loads where effects such as contact resistances are important and AC loads, where they lose their influence and the superconducting properties of the cable are mostly responsible for the physical behaviour. For DC loads, two methods to investigate the cable behaviour are presented: the analytic model which simply disregards the interaction of the superconductors via the magnetic field and hence is not able to account for self-field effects and a fully three-dimensional numerical model that is able to account for the self-field effects as well.

It is considerably easier financially and experimentally to measure short samples. In these, the contact resistances are large compared to the resistances of the superconducting tapes, even at high loads. Since the contact resistances and the tapes are connected like serial resistors, the contact resistances are therefore expected to have a big influence on the results in DC measurements, see Fig. 7.5.

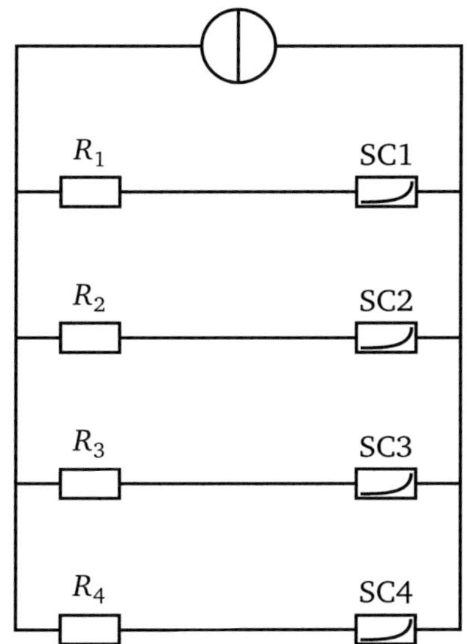

Figure 7.5: Circuit model of a TSTC cable with contact resistances.

In order to validate the various models, they will be compared with measurements of current distributions and among themselves. Afterwards, AC load cases will be considered that are also very interesting for long length cables as they suffer most from high hysteretic losses.

7.1 DC Behaviour of the TSTC Cable

The response of superconductors when loaded with very slow ramps and similar quasi-static excitations is independent of the flux relaxation effect. The superconductor merely acts as a resistor with no impedance. Thus, even superconducting systems may be considered with a conventional analytic model under these boundary conditions. Subsequently, the measurements of a four tape TSTC cable is shown and its behaviour will be modelled with the analytic model. The numerical model will also be compared to the analytic model. Due to the complexity of the measurements with the high number of parameters and the high computational effort required, the numerical model will be verified with a specific test case only.

The measurements include influences of the critical current of each tape, the n_p-value and the contact resistances all of which are difficult to measure and are therefore not necessarily precise and correct.

7.1.1 DC Transport Current Measurements

Discerning the electrical currents flowing in each tape is not possible using the simple four point measuring technique. The resistance of the superconductor is infinitesimal which is why the four point measuring technique cannot register any voltage. One possibility is to use hall probe sensors detecting the magnetic field around each conductor and calculating the electric current. This requires measuring as much of the magnetic field as possible which is achieved by surrounding each individual superconductor with a ferromagnetic ring that surrounds it as completely as possible [TCBM12].

Two configurations were measured: a TSTC cable with the tapes almost directly on top of each other, resulting in a gap space between two adjacent tapes of only 100 μm and the same cable arranged with insulating spacers, resulting in a gap of 4 mm. The measurements of the 80 cm long TSTC cable with the just mentioned separations are plotted in Fig. 7.7. Experimental measurement data was obtained during a research stay at MIT and collaboration with Dr. Takayasu who also published some of the results in [TCBM12].

The measurements are difficult to interpret as it is conceivable that some of the parameters such as contact resistances changed between experiments. Current sharing in the joints may also occur. The current distribution will be further analysed in Cap. 7.2.1.3.

The contact resistances were measured separately after the current distribution measurements and were determined to be roughly in the same range with the exception of the second coated conductor tape which was found to have a higher resistance [TCBM12]. This is also visible in the measurements as the second tape initially carries considerably less current than the three others. The contact resistances are listed in Tab. 7.1.

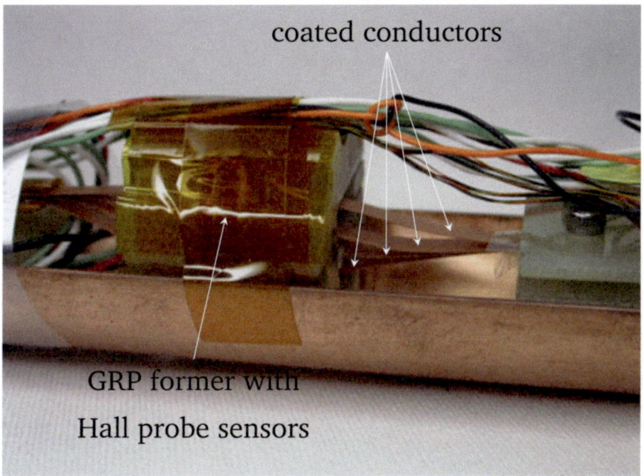

Figure 7.6: The measurement system for the four tape TSTC cable tape current measurements. The four-current-sensor assembly with four magnetic cores (11 mm × 6.4 mm × 3 mm) with Hall sensors is fixed on with some adhesive tape. Image courtesy of M. Takayasu, MIT. Also refer to [TCBM12].

TSTC cable contact resistances	
Tape Number	Resistance / $\mu\Omega$
R_1	0.67
R_2	1.06
R_3	0.65
R_4	0.73

Table 7.1: TSTC cable contact resistances.

7.1.2 Analytic DC Model

The experimental data showed curious behaviour of the TSTC cable not only with respect to the spacing but also considering one cable by itself. When completely disregarding the contact resistances some sort of symmetric behaviour could be expected. Since the contrary behaviour is observed, the behaviour seems largely governed by the contact resistances.

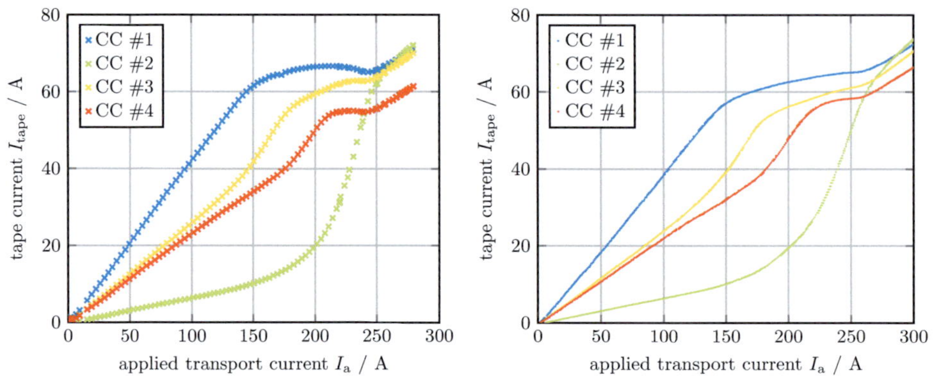

Figure 7.7: The graphic shows the experimental measurements of the TSTC cable for two configurations: with the tapes separated only $100\,\mu$m from each other (to the left) and with the tapes separated by spacers of 4 mm width (to the right). From the simulations with applied transport current, the tape current spread due to the tape spacing was expected to be more pronounced. In the measurements, the effect is barely visible. Data from [TCBM12] and from private communication with Dr. Takayasu.

Assuming quasi-static behaviour, we make use of the power-law to solve the following equation system:

$$
\begin{aligned}
I_1 &> 0 & V &= E_c \cdot (I_1/I_{c,1})^{n_1} + R_1 I_1 \\
I_2 &> 0 & V &= E_c \cdot (I_2/I_{c,2})^{n_2} + R_2 I_2 \\
I_3 &> 0 & V &= E_c \cdot (I_3/I_{c,3})^{n_3} + R_3 I_3 \\
I_4 &> 0 & V &= E_c \cdot (I_4/I_{c,4})^{n_4} + R_4 I_4.
\end{aligned}
\tag{7.1}
$$

Note that the voltage is actually the voltage drop over the cable so the units are in $\mathrm{V\,m^{-1}}$ and R_i is defined in $\Omega\,\mathrm{m^{-1}}$.

For the time being all tape critical currents and n_p-values are assumed identical. The influence of various contact resistance configurations is investigated. The conceivably simplest configuration is that all contact resistances are equally high:

$$
R_1 = R_2 = R_3 = R_4,
\tag{7.2}
$$

another assumes three contact resistances to be equal while one is different:

$$
R_1 = R_2 = R_3 = 2 \cdot R_4,
\tag{7.3}
$$

yet another has two contact resistances of equal magnitude while the other two are twice as large:

$$R_1 = 2 \cdot R_2 = 2 \cdot R_3 = R_4, \tag{7.4}$$

and lastly, we use the contact resistances listed in Tab. 7.1.

To compare the experimental results with the analytical model, the measured values for the critical currents, the n_p values and the contact resistances are used in the calculation:

$$
\begin{aligned}
I_{c,1} &= 82\ \text{A} & n_{p,1} &= 21 \\
I_{c,2} &= 82\ \text{A} & n_{p,2} &= 29 \\
I_{c,3} &= 84\ \text{A} & n_{p,3} &= 26 \\
I_{c,4} &= 83\ \text{A} & n_{p,4} &= 24.
\end{aligned}
\tag{7.5}
$$

The eight analytical simulations are seen in Fig. 7.8 for the case of equal critical current and power-law exponent n_p and in Fig. 7.9 for the measured values.

The influence of the contact resistance is visible in the linear regime of the plots. Above a threshold depending on the magnitude of the contact resistances, the power-law begins to dominate the formulas and the influence of the contact resistances begins to wane. In our simulations this is observable in the range between 200 A and 300 A.

The same behaviour is visible in the experiments, see Fig. 7.7. Since the linear regime ends much earlier in the plot showing experimental results, it is probable that the contact resistances were measured to be much higher than in reality. The measurement of contact resistances for superconductors is difficult and the contact resistance is dependent on environmental parameters such as temperature [Hua89]. On top of the difficult measurement, the sample was cut and its properties thus significantly altered [TCBM12].

The analytic macroscopic model provides a useful tool for investigating the behaviour of the TSTC cable in DC applications even though it is unable to account for the self field effect.

The higher the frequency, the less important the contact resistances become since the inductive behaviour of the cable and the self-field effect gain importance. Also, connections can be made with less than 10 nΩ resistance so that current redistribution due to contact resistance becomes even less important [Bar13, p. 106].

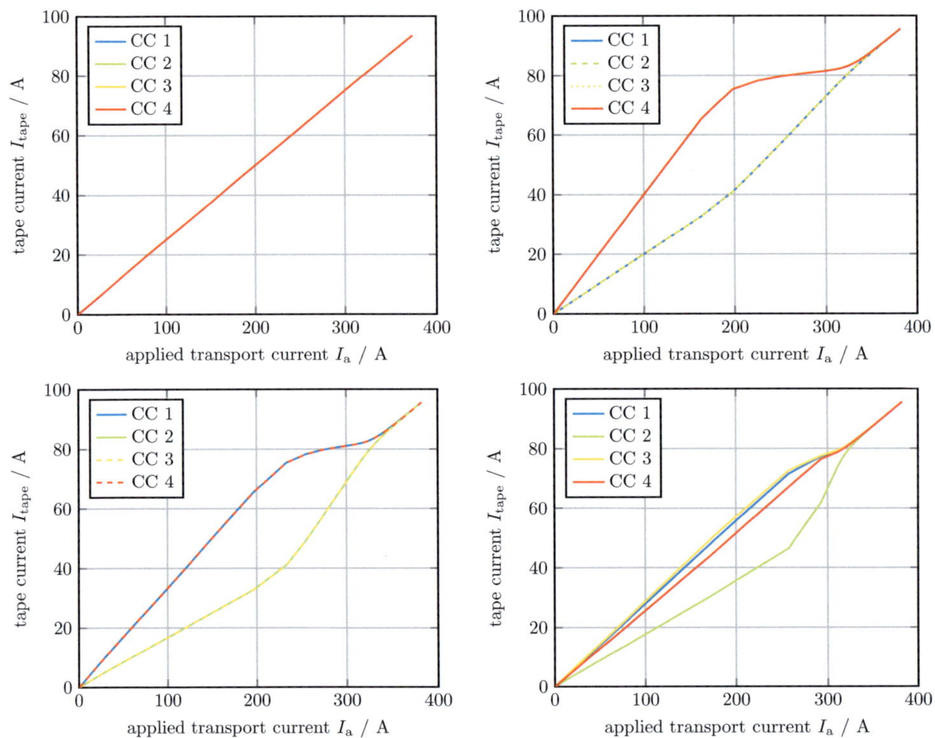

Figure 7.8: The four plots show analytical solutions of various contact resistance configurations for idealised conductors. The four coated conductors have the same critical current and the same power-law exponent. In the top left plot, all contact resistances are equal: $R_1 = R_2 = R_3 = R_4$, top right shows a configuration with $R_1 = R_2 = R_3 = 2 \cdot R_4$, bottom left shows $R_1 = 2 \cdot R_2 = 2 \cdot R_3 = R_4$ and bottom right shows $R_1 = 670\,\text{n}\Omega$, $R_2 = 1.06\,\mu\Omega$, $R_3 = 650\,\text{n}\Omega$, $R_4 = 730\,\text{n}\Omega$.

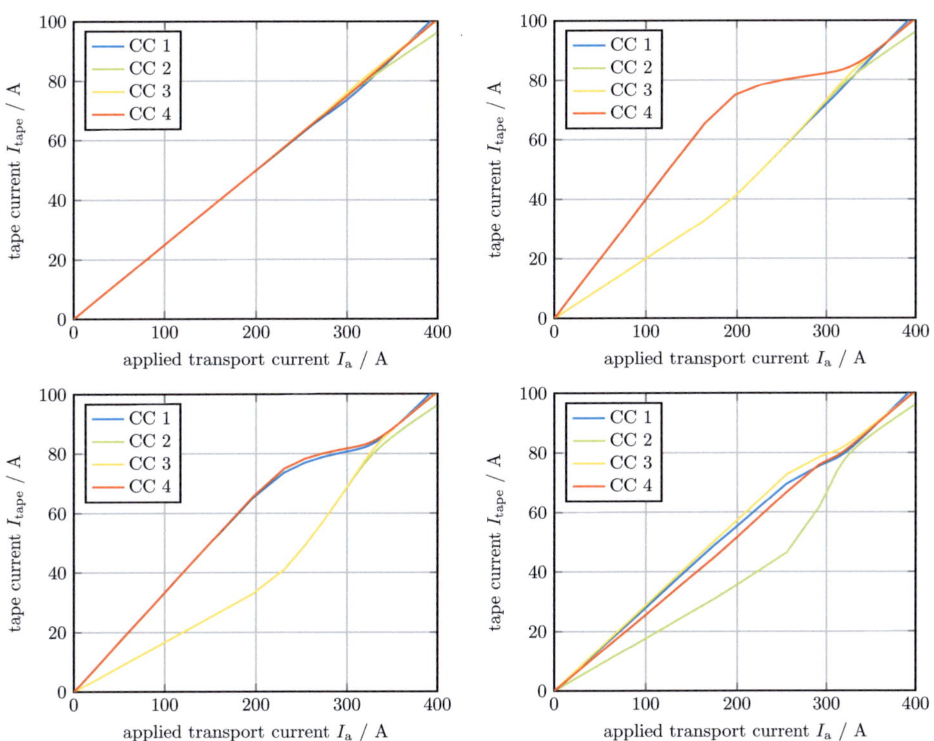

Figure 7.9: The four plots show analytical solutions of various contact resistance configurations using measured conductor properties. The critical tape currents of the four coated conductors are: $I_{c,1} = 82$ A, $I_{c,2} = 82$ A, $I_{c,3} = 84$ A, $I_{c,4} = 83$ A and their power-law exponents are: $n_{p,1} = 21$, $n_{p,2} = 29$, $n_{p,3} = 26$, $n_{p,4} = 24$. In the top left plot, all contact resistances are equal: $R_1 = R_2 = R_3 = R_4$, top right shows a configuration with $R_1 = R_2 = R_3 = 2 \cdot R_4$, bottom left shows $R_1 = 2 \cdot R_2 = 2 \cdot R_3 = R_4$ and bottom right shows $R_1 = 670$ nΩ, $R_2 = 1.06$ $\mu\Omega$, $R_3 = 650$ nΩ, $R_4 = 730$ nΩ.

7.1.3 Fitting the Measurements with the Analytic Model

Assuming the electric transport current measurements were slow enough to be considered quasi-static, it is permissible to use the analytic model in order to match the measurement data. This allows the determination of the tape parameters like critical current and contact resistances. The measurements of the Twisted Stacked Tape Conductor cable with 4 mm tape spacing are used since in these the self-field effect is smaller than in the tightly stacked cable. The result of the fit of the analytical model to the measured data is shown in Fig. 7.10.

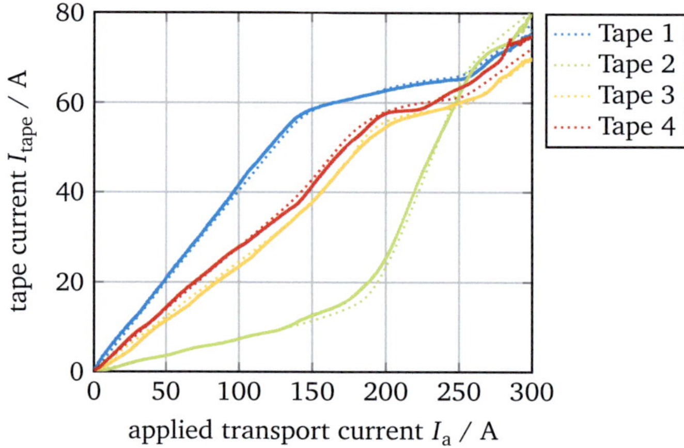

Figure 7.10: The TSTC cable current measurements (thick lines) are compared with the analytic model (dotted lines).

In order to successfully fit the measurements, data processing was used to clean up the measurement data. Especially towards very low applied electric transport currents, the data was assumed to be inaccurate. The following procedure was conceived by Dr. Zermeno [Zn13, private communication].

The tape current data points above 10 A and below 100 A (for tapes 2, 3 and 4) and below 80 A (for tape 1) of applied transport current were used in order to determine a current offset using a linear regression analysis of the linear part of the current profiles. The current offset was then subtracted from the measurements. Afterwards, the contact resistances relative to an arbitrarily chosen contact resistance are calculated (the current distribution is initially only dependent on the relation of the contact resistances with respect to each other). The contact resistance of the fourth tape was assumed to be $1\,\mu\Omega$ which yields the following tape contact resistances:

$$
\begin{aligned}
R_1 &= 0.67\,\mu\Omega & R_2 &= 3.78\,\mu\Omega \\
R_3 &= 1.13\,\mu\Omega & R_4 &= 1\,\mu\Omega.
\end{aligned}
\tag{7.6}
$$

Assuming the same n_p-value of 35 in all the tapes for the sake of simplicity and iteratively adjusting the critical current values yields the following critical current values for the coated conductor tapes:

$$
\begin{aligned}
I_{c,1} &= 64.5\,\text{A} & I_{c,2} &= 67\,\text{A} \\
I_{c,3} &= 58\,\text{A} & I_{c,4} &= 60\,\text{A}.
\end{aligned}
\tag{7.7}
$$

The values tabulated in Eq. 7.6 and in Eq. 7.7 result in the plot shown in Fig. 7.10. The extracted contact resistances are very reliable as they are directly connected to the linear slope of the measurement data for each tape. The direct measurements of the contact resistances were conducted after the tape was cut which most likely altered the properties somewhat. The ratio of the contact resistances determines the current distribution. This means the contact resistances are free to be scaled in magnitude, if only the correct ratio is kept. The critical currents are not freely rescalable however. The tape behaviour is almost perfectly reproduced by the analytic model.

The difference in measured and calculated parameter values is probably a combination of various influences. For example, the tape critical currents in the assembled cable are different from those of the isolated tapes due to the self-field effect and possibly also thermal reasons. Contact resistances on the other hand are generally difficult to measure as they may change when the cable is handled in order to conduct the measurements.

7.1.4 Three-Dimensional FEM Model

The three-dimensional model was built to account for the twisted geometry of the TSTC model. A stack of twisted coated conductors, see Fig. 7.11, is surrounded by a larger air region, see Fig. 7.12. In order to successfully mesh the geometry, a manual twisting procedure is required. This is why the geometry shows sectioning even without being meshed.

Comparison of tape parameters of a TSTC cable			
Tape Number	Tape Critical Current / A		Contact Resistance / $\mu\Omega$	
	measured	simulated	measured	extracted
1	82	64.5	0.67	0.67
2	82	67	1.06	3.78
3	84	58	0.65	1.13
4	83	60	0.73	1

Table 7.2: Coated conductor tape parameter comparison between direct measurements and extracted using analytic simulation.

The contact resistances are modelled as distinctive domains disconnected from the primary simulation environment to prevent unwanted interaction, see Cap. 3.2.6. They are surrounded by an air domain of their own, see Fig. 7.13. The four massive blocks are the contact resistances.

The transport current is fed into one side of all of the four contact resistance blocks concurrently and then one continuity boundary condition of the kind explained in Cap. 3.2.6 feeds the current from the other side of each contact resistance block to the corresponding tape. One block and one tape belonging together have been coloured red to illustrate the concept. The tapes have additionally been separated in order to be better visible.

When calculating the correct resistivity for the contact resistance domains one has to be mindful of the skin effect as mentioned earlier in Cap. 3.2.6. If the skin effect leads to a current distribution that is smaller than the dimensions of the contact domain, the domain has either to be made smaller or to be meshed more densely to ensure the correct application of the effective resistivity.

The discretisation of the simulation environment is vitally important to the efficiency and speed of numerical FEM simulations. Great care was therefore taken to ensure a high quality mesh. Defining the term "high-quality" is difficult, but as a general guideline, high-quality meshes look aesthetic and have as few elements as possible. For the coated conductor tapes, only one element was used for the thickness. If not using field dependent critical current densities, this is an admissible simplification. The tapes were also, like in the two-dimensional model, represented by quadrilateral elements. This effectively halves the amount of elements needed while actually improving the simulation accuracy. The air region was meshed with prismatic elements.

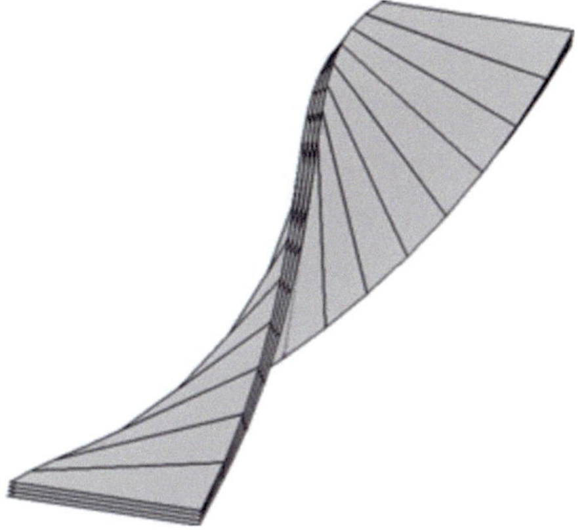

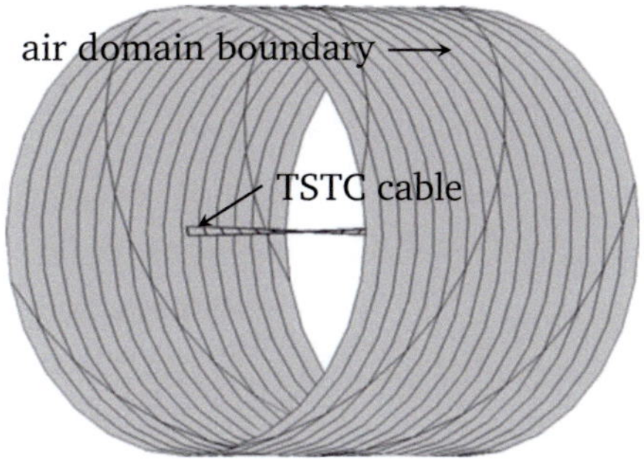

Figure 7.11: View along a stack of twisted coated conductors in the three-dimensional FEM model. The subsections of each tape are required for the discretisation.

Figure 7.12: View of a stack of twisted coated conductors surrounded by the air domain in the three-dimensional FEM model. The axial faces of the air domain are hidden to allow a view of the coated conductor stack situated in the middle.

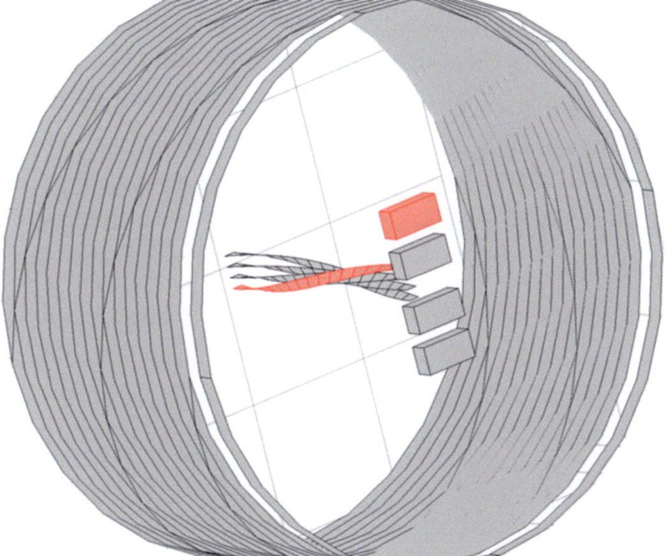

Figure 7.13: View of the three-dimensional TSTC model with contact resistances. The air region is again mostly hidden and only the outer boundaries are shown. The four massive blocks are the contact resistances. The transport current is fed into all of the four contact resistance blocks concurrently and then one continuity boundary condition feeds the current from the other side of each contact resistance block to the corresponding tape. One block and one tape belonging together have been coloured red. The tapes have additionally been separated in order to be better visible.

Since using the trapezoidal elements to mesh the tape and the prismatic elements to mesh the air region leads to identical cross-sections along the transposition, the simulation is quite fast. Another advantage is to be able to impose periodicity (or mirror) boundary conditions on the initial and the final faces, also see [ZnGS13]. This is only possible for a perfect identity map. In other words, the simulation field components at each node on one face have to be mirrored to a corresponding node on the other face which is only possible if the meshes on both faces are identical. If the meshes are not identical, this mapping is only possible approximately which leads to problems with the convergence. Also, it is not possible in COMSOL.

7.1.5 DC Compliance of the Three-Dimensional FEM Model

In the extreme case of quasi static loads, the analytic model provides a complete description of an ideal superconducting cable by using the simple formulation in Cap. 7.1.2. As such it can be used to validate the three-dimensional model when static loads are assumed. For this purpose, a simple two tape configuration is considered. The contact resistances of the two tapes differ: one tape has a contact resistance of 10 % of the resistance of the coated conductor tape at I_c while the other tape has a contact resistance of 20 % I_c. The tapes are 1 cm wide and 1 μm thick with a critical current of 300 A and an n_p-value of 35. Their spacing should not matter as the simulation is running in quasi-DC with an extremely low ramp rate; the spacing of the tapes is 1 cm.

The qualitative and quantitative agreement of the two models under DC load is excellent, see Fig. 7.14. For DC cases, the much faster analytic model can be used. The striking agreement of the three-dimensional model with the analytic model emphasises its reliability. It may be used for all those cases where the analytic model is not applicable.

7.2 AC Behaviour of the TSTC Cable

When the behaviour of the Twisted Stacked Tape Conductor cable under sufficiently rapid transient electric transport currents is under investigation or when background magnetic fields, either static or dynamic, have to be simulated the three-dimensional model has to be used. Combinations of the various loads are also possible. The numerical model has to be used since the analytic model is not able to account for the inductance in the superconductors.

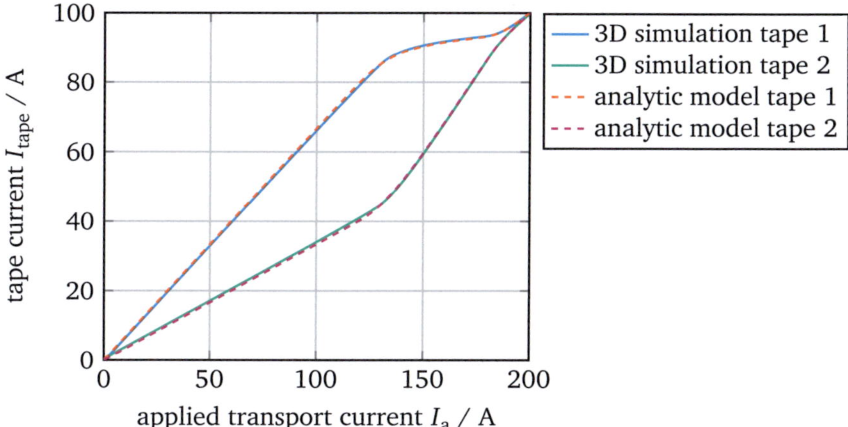

Figure 7.14: The current distribution of the analytic and the numeric TSTC model for a two tape cable is shown as a function of applied transport current. Under DC load, the agreement is excellent.

7.2.1 Influence of Self-Field Effect

The self-field effect should affect the tape current distribution in a Twisted Stacked Tape Conductor cable even under quasi-static load when the contact resistances of the tapes are disregarded entirely. The three cases of applied transport current, background magnetic field and lastly the influence of the tape spacing under applied transport current are explored.

7.2.1.1 Applying Background Magnetic Field

A global background magnetic field of three times the local critical field (see the accompanying definitions of Eq. 3.9; in this instance amounts to 89 mT) was imposed as an excitatory influence on a twisted stacked tape conductor cable consisting of four coated conductors. The superconductors were 4 mm wide, 10 μm thick and had a critical current of 300 A and an n_{p}-value of 35. While the I_{c} value may seem unrealistically high for a 4 mm tape, it is worthwhile keeping in mind the high thickness which results in a critical current density of $7.5 \cdot 10^9$ A m^{-2} which is realistic. The high thickness reduces computational effort and speeds up the simulation. The critical electric field was set to 1 μV cm^{-1} and the duration of the linearly ramped simulation was 100 ms. The twist pitch of the cable was 20 cm.

Because of the twisted design of the cable, a background magnetic field must not lead to a net current as the shielding currents induced in the cable effectively cancel each other out. Plotting the current in each tape as a function of the time of the simulation, this behaviour is indeed observed, see Fig. 7.15. The currents in tape one and four and in tape two and three, respectively, have the same amplitude but opposing polarity.

The applied magnetic background field results in shielding currents flowing that form loops in each twist pitch. The loops close where the magnetic field is parallel to the face of the superconducting tapes and crosses from one lateral edge to the other.

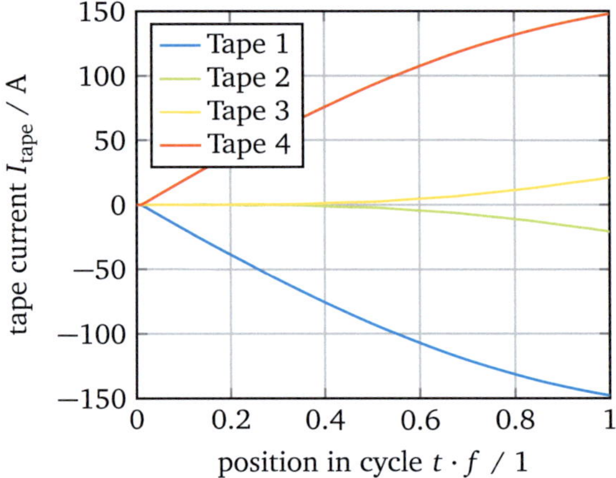

Figure 7.15: Simulation of a TSTC cable with the individual tape currents plotted versus time in the cycle with the background magnetic field being linearly ramped up. At the last timestep of the simulation, the background magnetic field reaches 89 mT ($3 \cdot H_c$). Initially, the two outer tapes carry most of the current thereby shielding the two inner tapes. Later when the outer tapes are close to saturation, the inner tapes pick up more current. Note the shielding currents flowing in opposing directions which results in zero net transport current.

The current distribution of the current component flowing along the axis of the tape is plotted in Fig. 7.16. As the TSTC cable is not saturated by the background magnetic field of $3 \cdot H_c$, the tapes carry differing currents. The outer tape shown on the top left carries considerably more current than the inner tape shown to its right. Only one half of the TSTC is plotted with two tapes showing. Since the spacing between them is only 100 μm they are perceived as one single tape. The outer tape shields the inner tape from the magnetic field and thus carries a higher current.

The central parts of the strip (halfway from one end of the simulation geometry to the other end) show no current flowing along the cable's axis. Here, the magnetic background field is parallel to the superconductor surface and the shielding currents loop back. Because the current is flowing in one direction on one side of the tapes and antiparallel on the other side, the net current is zero as is expected since no electric transport current should appear in the background magnetic field case as only shielding currents are flowing.

Each of the two outer and the two inner tapes behave exactly alike except that their behaviour is chiral. With this knowledge, it is sufficient to regard only two coated conductor tapes in the analysis since the behaviour of the two others is the same.

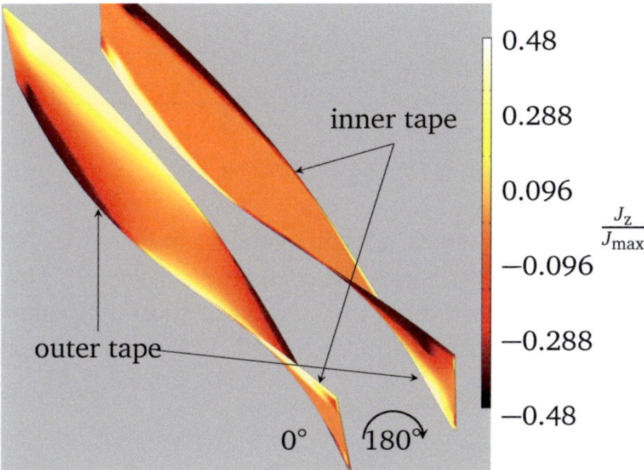

Figure 7.16: Plotted is the current component flowing along the axis of the TSTC cable when the latter is subjected to a background magnetic field. The top left tape is one of the outer tapes, the one to the right is one of the inner tapes. Each of the two structures shows two tapes: one outer and one inner tape. The right structure is rotated by $180°$ so the second tape is visible. The TSTC cable is not saturated by the relatively small background field of $3 \cdot H_c$. Both tapes carry different currents therefore and it is obvious that substantially more current is flowing in the outer tape. Refer to Fig. 7.15 for TSTC properties.

7.2.1.2 Applying Electric Transport Current

The same Twisted Stacked Tape Conductor cable that has been simulated in the applied magnetic field investigation is also used when applying electric transport current.

The applied boundary condition is that of a linear ramp of transport current that reaches the critical current at the end of the ramp. Opposed to the case of background magnetic field, the total net current is different from zero and equal to the sum of all tape currents. The current in each tape will however not necessarily be the same.

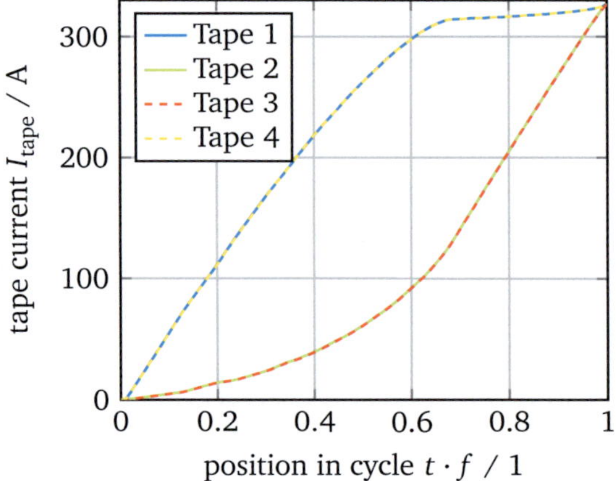

Figure 7.17: Individual tape current plotted versus time in the cycle. At the last timestep of the simulation, the current has reached 100 % of the critical current. Initially, the two outer tapes carry most of the current thereby shielding the two inner tapes. Later when the outer tapes are saturated, the inner tapes begin to pick up more of the current.

Indeed, solely due to the self-field effect, the tape currents are spread apart as may be observed in Fig. 7.17. Unlike in the case of the measurements and the DC simulations mentioned in Cap. 7.1.2, this difference in tape currents is not due to contact resistances. The central tapes are shielded by the outer tapes so the inner ones carry current only at higher load factors when the outer tapes saturate. Each of the two outer and the two inner tapes behaves alike. This would be different if the tape parameters were different. Unlike in the case of background magnetic field, there is no chirality and the tapes carry equal current.

Since the tape current plotted in Fig. 7.17 is an integral observable which complicates understanding, it might be enlightening to again consider three-dimensional representations of the current distribution, see Fig. 7.18.

7.2.1.3 The Coated Conductor Tape Spacing

The distance between two coated conductor tapes in the Twisted Stacked Tape Conductor cable cannot have any influence in the analytic results presented in Cap. 7.1.2 since the self field effect is not taken into account in the analytic model presented.

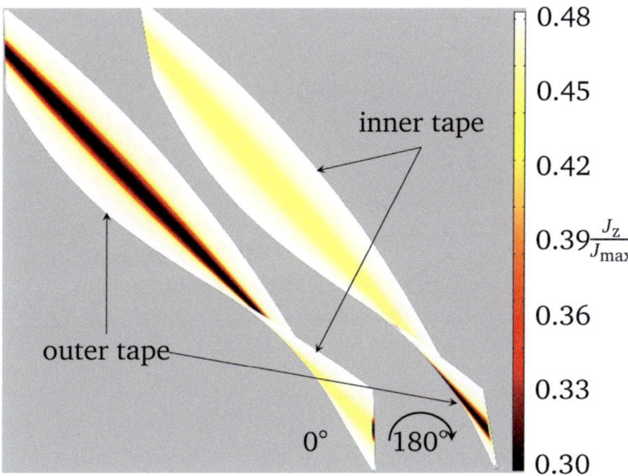

Figure 7.18: Plotted is the current component flowing along the axis of the TSTC cable when the latter is subjected to an electric transport current. The top left tape is one of the outer tapes, the one to the right is one of the inner tapes. Each of the two structures shows two tapes: one outer and one inner tape. The right structure is rotated by $180°$ so the second tape is visible. The TSTC cable is not saturated by the electric transport current yet. The configuration shown is about 90 % of I_c. Both tapes carry different currents therefore and again the outer tape is saturated earlier while the inner tape shares a reduced fraction of the load. Refer to Fig. 7.15 for TSTC properties.

The three-dimensional FEM simulations however allow for an investigation of this effect.

Looking at the experimental measurements in Fig. 7.6, the difference between the two cases is not as large as might be expected from the simulations presented in Cap. 7.2.1.2. It was already established that DC measurements are mainly governed by the coated conductor tape contact resistances. For AC applications on the other hand, the influence of the tape spacing is of interest. The contact resistances are therefore assumed zero and two simulations only differing in the gap distance between two adjacent coated conductors are compared, see Fig. 7.19.

The difference in current distributions visible in Fig. 7.19 due to the different gap size between two adjacent superconducting tapes is quite pronounced. The spread between the current distribution is reduced in the 4 mm configuration as expected since the self field effect has less influence. Also, the spread is maintained until higher transport currents until the normal resistance of the superconductors dominates. At high currents, all tapes are limited by their critical transport current.

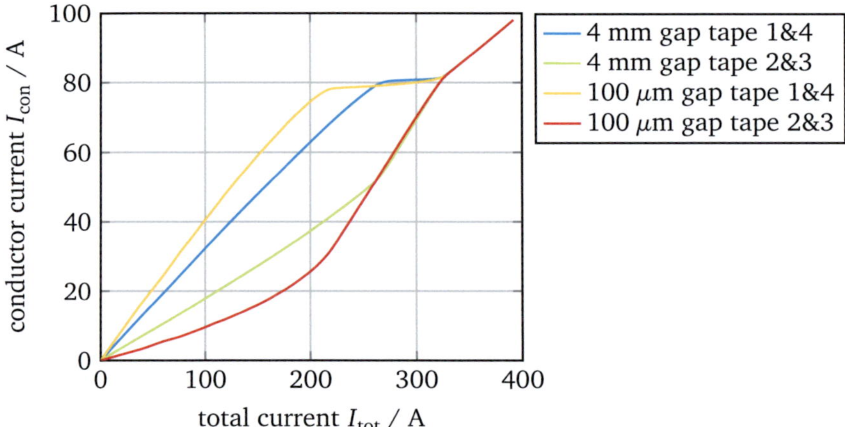

Figure 7.19: The currents flowing in each tape are plotted versus the total applied electric transport current. Because contact resistances are not taken into account, coated conductor tape one and four as well as two and three are showing identical behaviour due to the symmetric tape layout resulting in only two visible curves per gap configuration.

7.2.2 Sinusoidal Loads

The load profiles considered up until now were either DC applications or slow ramps which are also considered quasi-static. These loads lead only to low hysteretic losses as the polarity of the electric transport current is not inverted as in electric power grid applications. Other applications requiring high current cables operate at higher frequencies making hysteretic loss investigations highly interesting.

In order to appraise the hysteretic losses, a sinusoidal transport current is applied to a quadruple coated conductor TSTC cable. The conductor properties are the same as those mentioned in Cap. 7.2.1.1: each of the tapes is 4 mm wide and 10 μm thick, has an n_p-value of 35 and a critical current density of 300 A. The sinusoidal current peaks at I_c during the cycle which has a duration of 20 ms. Due to the high frequency, contact resistances are omitted since the behaviour would mostly be governed by the superconductor behaviour anyway.

7.2.2.1 Current Distributions

The resulting current distribution is a response to the complicated interaction of the tapes and varies in each tape and with time, see Fig. 7.20. Just as before, the inner tapes carry less current and the current profile shows the load profile lagging behind that of the outer tapes. This behaviour is well known from previous investigations as the outer conductors shield the inner ones. Contrary to the case of slow ramps where the response of the tapes is in sync and the difference in tape current more pronounced, the amplitude difference in tape currents between the outer and inner tapes is much smaller.

The current profile of the inner tapes shows a distinctly delayed and slightly dispersed response. The outer tapes on the other hand lead the applied transport current while showing steeper transitions. The fact that the phase-shift of the outer tapes' current profile is negative is interesting and important when considering hysteretic losses.

Opposed to the current distribution in a Röbel cable where due to the complete transposition of the tapes, every tape carries the same current [GP10], in the TSTC cable, considerably higher load is carried by the outer tapes.

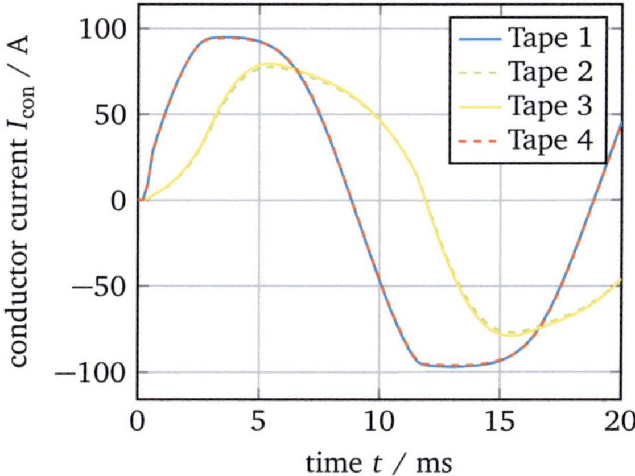

Figure 7.20: Current distribution as a function of time in a TSTC cable with four coated conductor tapes (4 mm wide and 10 μm thick with a critical current of 300 A and an n_p-value of 35) due to sinusoidal applied electric transport current. Note the delayed response of the inner tapes with the tape current in the central tapes lagging behind that of the outer tapes due to shielding effects. The difference in current load is much smaller in the sinusoidal than in the ramped case.

7.2.2.2 Hysteretic Losses

The hysteretic losses resulting from the current distribution plotted in Fig. 7.20 is shown in Fig. 7.21. The losses in the outer tapes are much higher than those in the inner tapes but interestingly and opposed to the behaviour of the current profiles, they are almost in sync, meaning there is only a small phase shift between the peaks. The instantaneous losses in the first half cycle show initial magnetisation effects and are therefore smaller than those in the second half cycle, see also Cap. 3.2.4.

The losses are higher in the outer tapes because the fraction of current carried by these tapes is also higher. If the coated conductor tapes were fully transposed, there would be no phase shift in the cable and hence the hysteretic losses would not only be the same in all the tapes but generally be lower. If the tape parameters like critical tape current and n_p-value are the same, the response of the two inner and the two outer tapes is the same (apart from small numerical errors). If however those parameters are different in each tape, the hysteretic loss response of each tape will be different as well. To which degree this discrepancy occurs depends on the extent to which the tape parameters differ. The effect of this influence depends strongly on the tape parameters respective to each other: the hysteretic losses could be exacerbated or diminished. Further investigations have to determine if this can be used to improve the tape performance.

When comparing the two figures just mentioned, it is enlightening to take notice of the fact that the hysteretic losses in all the tapes peak between the point where the electric current reaches its maximum in the outer tapes and the point where the electric current reaches its maximum in the inner tapes. In other words: the phase shift in the tapes leads to increased losses which is why a Röbel cable performs so well in AC applications. One way to reduce hysteretic losses in the TSTC therefore is to minimise the phase-shift. Another way to reduce hysteretic losses is to apply the concept of ferromagnetic shielding. Both these options have to be explored in future investigations.

7.3 Further Development of the TSTC Cable and Model

The tape layout discussed so far is only one possibility of arranging coated conductors in a twisted structure. Additionally, for very high current applications, multiple TSTC cables have to be combined. The twisted stacks are soldered into cylindrical formers in order for the assembly to be more mechanically robust and are then combined into cable assemblies. Fig. 7.22 shows the cross-sections of three configurations as well as an alternate method of stacking and twisting the tapes. Instead of one TSTC cable consisting of only a single twisted stack, a cylindrical metal former characterised by possessing three helical grooves is used into which three stacks of coated conductors are inserted. The three stacks form a twisted helix. Theoretically, this configuration

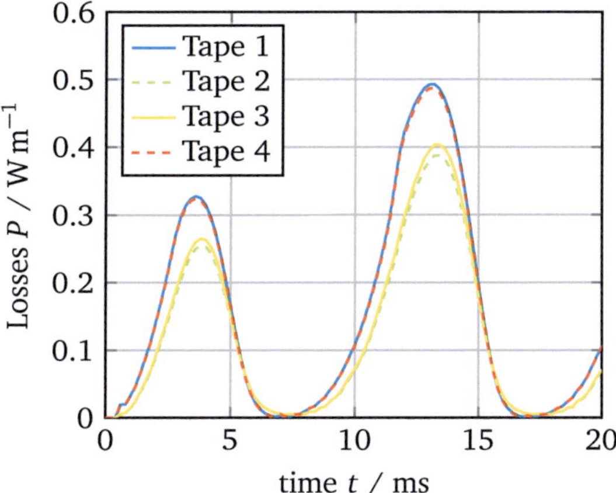

Figure 7.21: Instantaneous losses plotted versus time in the tapes of a TSTC during application of a sinusoidal electric transport current. The two symmetrically positioned tapes do not behave identically due to their physical properties being different. See Cap. 7.1.2 for their critical current and n_p values.

could also be extended to contain an arbitrary amount of grooves and thus, stacks. Beneficial effects are expected if a way could be found to transpose single coated conductor tapes from the bottom of one stack to the top of another in order to obtain a better current distribution.

The coaligned triple helical configuration is much easier fitted with ferromagnetic shielding which is why further interest is warranted in this design especially.

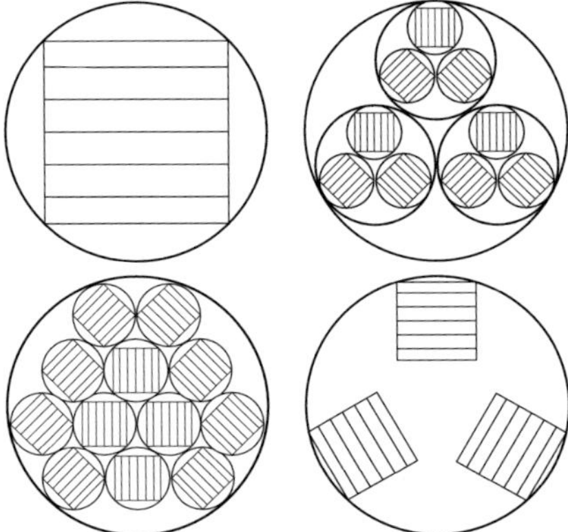

Figure 7.22: Cross-sectional drawings of various Twisted Stacked Tape Conductor cable configurations. Top left: single TSTC, top right: 3x3 conductor layout, bottom left: 12 conductor layout, bottom right: coaligned rotated twisted stacked tape conductor. The last configuration shown at the lower right consists of three stacks of coated conductors instead of just one stack that are inserted into grooves. Multiple of these coaligned TSTC assemblies can then again be combined to form larger cables as those shown at to the upper right and lower left.

8 Conclusions and Perspectives

This work provides researchers and developers alike with tools to investigate and predict hysteretic losses in superconducting applications while at the same time proving the feasibility of ferromagnetic shielding. Besides a two-dimensional H-model implementation with the ability to model ferromagnetic materials and conduct multiphysics analysis, a three-dimensional implementation for the simulation of magnetic fields and current distributions in superconductors was built. The model presented in this work is one of two in existence worldwide. Different descriptions for superconducting and ferromagnetic materials such as field- and strain- or temperature-dependency were also implemented. The model was verified to reproduce analytical predictions. Experimental work complemented the verification process and coated conductor samples with and without ferromagnetic shields were measured in both background magnetic field as well as electric transport current measurements.

During the experimental part of this thesis, a method for coating the coated conductors with ferromagnetic shielding using electroplating was developed and multiple samples were prepared. The first batch of ten samples consisted of silver stabilised coated conductors that were plated with a nickel layer of 20 μm. The second batch of ten samples differed from the first because of an additional copper sheet around the coated conductor. The copper protected the superconducting layer during the electroplating procedure. The silver layer was still present to forestall copper poisoning of the superconducting layer. The concept of ferromagnetic shielding has been proposed before, but nobody had investigated coated conductor tapes before and no experimental data was available until the experiments published in this thesis.

In the experimental section this thesis demonstrates the potential of ferromagnetic shielding to reduce hysteretic losses in heterostructures of ferromagnets and super-conductors. When considering single coated conductors, this reduction is up to 60 % in the magnetic background field case and up to 50 % in the applied transport current case. For stacks of coated conductor tapes, the hysteretic losses are reduced by up to 50 %, both with transport current as well as in background field. Bifilar coils even show a reduction of the hysteretic losses of one order of magnitude, meaning the losses with shields are as low as 10 % of the design without shields. Pancake coils profit slightly less from ferromagnetic shielding, the highest achievable reduction of hysteretic losses is about 40 %.

With increasing geometric and physical complexity, numerical models may become slow. An important task of this thesis was therefore besides the implementation, verification and the investigation of ferromagnetic shielding geometries the optimisation of the numerical model. In this instance, the optimisation included finding a more elegant and computationally less expensive way to implement the material laws and governing equations but also reducing the mesh complexity without decreasing the modelling accuracy. Optimisation also means exploiting and implementing symmetry and boundary conditions resulting therefrom. For simple coated conductor geometries, this optimisation resulted in simulations requiring about two minutes instead of one hour. For the three-dimensional model, the increase in computational efficiency resulted in simulations requiring four hours instead of two weeks. For the former case mentioned which regards the two-dimensional model, the savings amount to roughly 96 % of the total computation time so that the solution only requires 4 % of the time. The optimisations for the three-dimensional model result in the optimised version requiring about 1 % of the time of the unoptimised version.

A comparatively accessible interface with control over all the simulation parameters was programmed in order to allow researchers with little experience to access and alter the numerical model. It is found in the appendix. The script allows applying arbitrarily complex load profiles, not only ramped electric current and background magnetic field or sinusoidal loads.

The three-dimensional numerical model is the first one for superconductor simulation that is capable of simulating contact resistances. It is possible to include these boundary conditions without affecting the rest of the simulation environment, resulting in unbroken translation symmetry. This new boundary condition can also be implemented in other models and constitutes an important contribution to the numerical modelling of superconductors.

Due to these achievements, the Twisted Stacked Tape Conductor cable could be modelled and the influence of the geometry and of the contact resistances was investigated. Contrary to the analytical model also presented which is valid only for dc applications, the numerical model is additionally capable of simulating ac loads. Concepts for reducing the hysteretic losses in the TSTC cable are proposed that could lead to a more competitive performance in ac applications.

With the work on the TSTC cable, the basis for a comprehensive comparison of all high current cable designs has been laid. A requirement for finding the optimal design for high current fusion application cables, which are for example required in order to power the plasma containment high magnetic field coils. Topological optimisation can further improve the cable performance and the application of ferromagnetic shielding in the cable designs is another very interesting possibility.

Understanding the behaviour of complex systems including isolators, conductors, superconductors and ferromagnets will help design and optimise appliances for increased efficiency and high power density. This thesis contributes a powerful tool to master those challenges, demonstrates the possibility of significantly optimising superconducting applications and paves the way for further research into the field of hysteretic loss reduction by ferromagnetic shielding.

9 Bibliography

[AJS⁺11]   Abrahamsen, A.B., Jensen, B.B., Seiler, E., Mijatovic, N., Rodriguez-Zermeño, V.M., Andersen, N.H. and Ostergard, J. "Feasibility study of 5MW superconducting wind turbine generator." *Physica C: Superconductivity* **471**, 1464–1469 (2011). http://dx.doi.org/10.1016/j.physc.2011.05.217. (Cited on page 1)

[AMBM98]   Amemiya, N., Murasawa, S.I., Banno, N. and Miyamoto, K. "Numerical modelings of superconducting wires for AC loss calculations." *Physica C: Superconductivity* **310**, 16–29 (1998). http://dx.doi.org/10.1016/S0921-4534(98)00427-4. (Cited on pages 4 and 42)

[AMS⁺09]   Abrahamsen, A.B., Mijatovic, N., Seiler, E., Sorensen, M.P., Koch, M., Norgard, P.B., Pedersen, N.F., Traeholt, C., Andersen, N.H. and Ostergard, J. "Design Study of 10 kW Superconducting Generator for Wind Turbine Applications." *IEEE Transactions on Applied Superconductivity* **19**, 1678–1682 (2009). http://dx.doi.org/10.1109/TASC.2009.2017697. (Cited on page 1)

[AMS⁺10]   Abrahamsen, A.B., Mijatovic, N., Seiler, E., Zirngibl, T., Traeholt, C., Norgaard, P.B., Pedersen, N.F., Andersen, N.H. and Ostergard, J. "Superconducting wind turbine generators." *Superconductor Science and Technology* **23**, 034019 (2010). http://dx.doi.org/10.1088/0953-2048/23/3/034019. (Cited on page 1)

[Bar13]   Barth, C. *High Temperature Superconductor Cable Concepts for Fusion Magnets.* KIT Scientific Publishing, Karlsruhe (2013). (Cited on pages 24, 25, 133, 135, 136 and 141)

[BCS57]   Bardeen, J., Cooper, L.N. and Schrieffer, J.R. "Theory of Superconductivity." *Physical Review* **108**, 1175–1204 (1957). http://dx.doi.org/10.1103/PhysRev.108.1175. (Cited on pages 6 and 7)

[BD62]   Bean, C.P. and Doyle, M.V. "Superconductors as Permanent Magnets." *Journal of Applied Physics* **33**, 3334–3337 (1962). (Cited on pages 2 and 33)

[Bea62]   Bean, C. "Magnetization of Hard Superconductors." *Physical Review Letters* **8**, 250–253 (1962). http://dx.doi.org/10.1103/PhysRevLett.8. 250. (Cited on pages 2, 33 and 35)

[Bea64]   Bean, C.P. "Magnetization of High-Field Superconductors." *Reviews of Modern Physics* **36**, 31–39 (1964). http://dx.doi.org/10.1103/ RevModPhys.36.31. (Cited on pages 2 and 33)

[BGL92]   Blatter, G., Geshkenbein, V.B. and Larkin, A.I. "From isotropic to anisotropic superconductors: A scaling approach." *Physical Review Letters* **68**, 875–878 (1992). http://dx.doi.org/10.1103/PhysRevLett. 68.875. (Cited on page 35)

[BGM07]   Brambilla, R., Grilli, F. and Martini, L. "Development of an edge-element model for AC loss computation of high-temperature supercon-ductors." *Superconductor Science and Technology* **20**, 16–24 (2007). http://dx.doi.org/10.1088/0953-2048/20/1/004. (Cited on pages 4, 42 and 60)

[BGN⁺09]   Brambilla, R., Grilli, F., Nguyen, D.N., Martini, L. and Sirois, F. "AC losses in thin superconductors: the integral equation method ap-plied to stacks and windings." *Superconductor Science and Technology* **22**, 075 018 (2009). http://dx.doi.org/10.1088/0953-2048/22/7/ 075018. (Cited on page 106)

[BI93]   Brandt, E.H. and Indenbom, M.V. "Type-II-superconductor strip with current in a perpendicular magnetic field." *Physical Review B* **48**, 12 893–12 906 (1993). http://dx.doi.org/10.1103/PhysRevB.48. 12893. (Cited on pages 2, 16, 37, 38, 78, 83 and 104)

[BK04]   Buckel, W. and Kleiner, R. *Superconductivity: Fundamentals and Ap-plications*. Wiley-VCH Verlag GmbH & Co. KGaA, Weinheim, 2nd ed. (2004). ISBN 3527403493. (Cited on pages 9 and 13)

[BL01]   Badía, A. and López, C. "Critical State Theory for Nonparallel Flux Line Lattices in Type-II Superconductors." *Physical Review Letters* **87**, 127 004 (2001). http://dx.doi.org/10.1103/PhysRevLett.87.127004. (Cited on page 49)

[BM86]   Bednorz, J.G. and Müller, K.A. "Possible High-T$_c$ Superconductivity in the Ba-La-Cu-O System." *Zeitschrift für Physik B Condensed Matter* **64**, 189–193 (1986). http://dx.doi.org/10.1007/BF01303701. (Cited on pages 1, 2 and 23)

[BMDH99]   Barnes, G., McCulloch, M.D. and Dew-Hughes, D. "Computer modelling of Type-II superconductors in applications." *Superconductor Science and Technology* **12**, 518–522 (1999). http://dx.doi.org/10.1088/0953-2048/12/8/308. (Cited on pages 4 and 42)

[BP83]   Boenig, H. and Paice, D. "Fault current limiter using a superconducting coil." *IEEE Transactions on Magnetics* **19**, 1051–1053 (1983). http://dx.doi.org/10.1109/TMAG.1983.1062396. (Cited on page 1)

[Bra96]   Brandt, E.H. "Superconductors of finite thickness in a perpendicular magnetic field: Strips and slabs." *Physical review. B, Condensed matter* **54**, 4246–4264 (1996). (Cited on page 2)

[Bra09]   Bray, J.W. "Superconductors in Applications; Some Practical Aspects." *IEEE Transactions on Applied Superconductivity* **19**, 2533–2539 (2009). http://dx.doi.org/10.1109/TASC.2009.2019287. (Cited on page 22)

[BSK99]   Bauer, M., Semerad, R. and Kinder, H. "YBCO films on metal substrates with biaxially aligned MgO buffer layers." *IEEE Transactions on Appiled Superconductivity* **9**, 1502–1505 (1999). http://dx.doi.org/10.1109/77.784678. (Cited on page 2)

[BTI+05]   Bohno, T., Tomioka, a., Imaizumi, M., Sanuki, Y., Yamamoto, T., Yasukawa, Y., Ono, H., Yagi, Y. and Iwadate, K. "Development of 66kV/6.9kV 2MVA prototype HTS power transformer." *Physica C: Superconductivity* **426-431**, 1402–1407 (2005). http://dx.doi.org/10.1016/j.physc.2005.03.080. (Cited on page 1)

[BV83]   Bossavit, A. and Vérité, J.C. "The T̈RIFOUC̈ode: Solving The 3-D Eddy-Currents Problem by using H as State Variable." *IEEE Trans. Magn.* **M**, 2465–2470 (1983). http://dx.doi.org/0018-9464/83/1100-2465. (Cited on pages 4 and 42)

[BW96]   Buntar, V. and Weber, H.W. "Magnetic properties of fullerene superconductors." *Superconductor Science and Technology* **9**, 599–615 (1996). http://dx.doi.org/10.1088/0953-2048/9/8/001. (Cited on page 20)

[Cam97]   Campbell, A.M. "Coupling losses in filamentary superconductors with a resistive barrier." *Superconductor Science and Technology* **10**, 932–935 (1997). http://dx.doi.org/10.1088/0953-2048/10/12/016. (Cited on page 20)

[Cam07]   Campbell, A.M. "A new method of determining the critical state in superconductors." *Superconductor Science and Technology* **20**, 292–295 (2007). http://dx.doi.org/10.1088/0953-2048/20/3/031. (Cited on pages 4 and 42)

[Cam09]   Campbell, A.M. "A direct method for obtaining the critical state in two and three dimensions." *Superconductor Science and Technology* **22**, 034 005 (2009). http://dx.doi.org/10.1088/0953-2048/22/3/034005. (Cited on pages 4 and 42)

[Cam11]   Campbell, A.M. "Flux cutting in superconductors." *Superconductor Science and Technology* **24**, 091 001 (2011). http://dx.doi.org/10.1088/0953-2048/24/9/091001. (Cited on page 49)

[CBK$^+$06] Cho, J., Bae, J.H., Kim, H.J., Sim, K.D., Kim, S., Jang, H.M., Lee, C.Y. and Kim, D.W. "Development of a single-phase 30m HTS power cable." *Cryogenics* **46**, 333–337 (2006). http://dx.doi.org/10.1016/j.cryogenics.2005.11.025. (Cited on pages 2 and 134)

[CFB$^+$01] Canfield, P., Finnemore, D., Bud'ko, S., Ostenson, J., Lapertot, G., Cunningham, C. and Petrovic, C. "Superconductivity in Dense MgB_2 Wires." *Physical Review Letters* **86**, 2423–2426 (2001). http://dx.doi.org/10.1103/PhysRevLett.86.2423. (Cited on page 22)

[CFD99]   Cordier, C., Flament, S. and Dubuc, C. "A 3-D finite element formulation for calculating Meissner currents in superconductors." *IEEE Transactions on Applied Superconductivity* **9**, 2–6 (1999). http://dx.doi.org/10.1109/77.763249. (Cited on pages 4 and 42)

[Cle08]   Clem, J.R. "Field and current distributions and ac losses in a bifilar stack of superconducting strips." *Physical Review B* **77**, 1–7 (2008). http://dx.doi.org/10.1103/PhysRevB.77.134506. (Cited on pages 114 and 115)

[CLY$^+$11] Cheng, L., Li, T., Yan, S., Sun, L., Yao, X. and Puzniak, R. "Large Size and High Performance of a Gd-Ba-Cu-O Bulk Superconductor Grown Using New Approaches." *Journal of the American Ceramic Society* **94**, 3139–3143 (2011). http://dx.doi.org/10.1111/j.1551-2916.2011.04497.x. (Cited on page 2)

[COM13]   COMSOL. "COMSOL Multiphysics." (2013). (Cited on page 43)

[Con]   Conectus. "Global Market for Superconductivity." (Cited on page 20)

[Coo56]   Cooper, L.N. "Bound Electron Pairs in a Degenerate Fermi Gas." *Physical Review* **104**, 1189–1190 (1956). http://dx.doi.org/10.1103/PhysRev.104.1189. (Cited on page 7)

[CR10]   Callister, W.D. and Rethwisch, D.G. *Materials Science and Engineering: An Introduction.* John Wiley & Sons, Inc., New York, 8th ed. (2010). ISBN 978-0-470-41997-7. (Cited on page 30)

[CWDC11]   Clem, J.R., Weigand, M., Durrell, J.H. and Campbell, A.M. "Theory and experiment testing flux-line cutting physics." *Superconductor Science and Technology* **24**, 062 002 (2011). http://dx.doi.org/10.1088/0953-2048/24/6/062002. (Cited on page 49)

[DBT$^+$04]   Durrell, J.H., Burnell, G., Tsaneva, V., Barber, Z., Blamire, M. and Evetts, J.E. "Critical currents in vicinal $YBa_2Cu_3O_{7-\delta}$ films." *Physical Review B* **70**, 214 508 (2004). http://dx.doi.org/10.1103/PhysRevB.70.214508. (Cited on page 35)

[DGL$^+$05]   Duckworth, R.C., Gouge, M.J., Lue, J.W., Thieme, C.L.H. and Verebelyi, D.T. "Substrate and Stabilization Effects on the Transport AC Losses in YBCO Coated Conductors." *IEEE Transactions on Applied Superconductivity* **15**, 1583–1586 (2005). http://dx.doi.org/10.1109/TASC.2005.849181. (Cited on page 29)

[DKE06]   DKE Deutsche Kommission Elektrotechnik Elektronik Informationstechnik im DIN und VDE. "IEC 61788-3 Ed.2: Superconductivity - Part 3: Critical current measurement - DC critical current of Ag- and/or Ag alloy-sheathed Bi-2212 and Bi-2223 oxide superconductors." (2006). (Cited on page 14)

[DKE07a]   DKE Deutsche Kommission Elektrotechnik Elektronik Informationstechnik im DIN und VDE. "IEC 61788-1 Supraleitfähigkeit - Teil 1: Messen des kritischen Stromes - Kritischer Strom (Gleichstrom) von Nb-Ti Verbundsupraleitern." (2007). (Cited on page 14)

[DKE07b]   DKE Deutsche Kommission Elektrotechnik Elektronik Informationstechnik im DIN und VDE. "IEC 61788-2 Supraleitfähigkeit - Teil 2: Messen des kritischen Stromes - Kritischer Strom (Gleichstrom) von Nb_3Sn Verbundsupraleitern." (2007). (Cited on page 14)

[DKH$^+$10]   Dommerque, R., Krämer, S., Hobl, A., Böhm, R., Bludau, M., Bock, J., Klaus, D., Piereder, H., Wilson, A., Krüger, T., Pfeiffer, G., Pfeiffer, K. and Elschner, S. "First commercial medium voltage superconducting fault-current limiters: production, test and installation." *Superconductor Science and Technology* **23**, 034020 (2010). http://dx.doi.org/10.1088/0953-2048/23/3/034020. (Cited on page 1)

[DN61]   Doll, R. and Näbauer, M. "Experimental Proof of Magnetic Flux Quantization in a Superconducting Ring." *Physical Review Letters* **7**, 51–52 (1961). http://dx.doi.org/10.1103/PhysRevLett.7.51. (Cited on page 12)

[DSJ+07]   Demko, J.A., Sauers, I., James, D.R., Gouge, M.J., Lindsay, D., Roden, M., Tolbert, J., Willén, D., Træ holt, C. and Nielsen, C.T. "Triaxial HTS Cable for the AEP Bixby Project." *IEEE Transactions on Applied Superconductivity* **17**, 2047–2050 (2007). http://dx.doi.org/http://dx.doi.org/10.1109/TASC.2007.897842. (Cited on pages 2 and 134)

[DTG+03]   Duckworth, R.C., Thompson, J.R., Gouge, M.J., Lue, J.W., Ijaduola, A.O., Yu, D. and Verebelyi, D.T. "Transport ac loss studies of YBCO coated conductors with nickel alloy substrates." *Superconductor Science and Technology* **16**, 1294–1298 (2003). (Cited on page 29)

[DYZ+06]   Dai, Y., Yan, L., Zhao, B., Song, S., Lei, Y. and Wang, Q. "Tests on a 6 T Conduction-Cooled Superconducting Magnet." *IEEE Transactions on Applied Superconductivity* **16**, 961–964 (2006). http://dx.doi.org/10.1109/TASC.2006.873331. (Cited on page 2)

[EKB+12]   Elschner, S., Kudymow, A., Brand, J., Fink, S., Goldacker, W., Grilli, F., Noe, M., Vojenciak, M., Hobl, A., Bludau, M., Jänke, C., Krämer, S. and Bock, J. "ENSYSTROB – Design, manufacturing and test of a 3-phase resistive fault current limiter based on coated conductors for medium voltage application." *Physica C: Superconductivity* **482**, 98–104 (2012). http://dx.doi.org/10.1016/j.physc.2012.04.025. (Cited on page 112)

[EKF+11]   Elschner, S., Kudymow, A., Fink, S., Goldacker, W., Grilli, F., Schacherer, C., Hobl, A., Bock, J. and Noe, M. "ENSYSTROB—Resistive Fault Current Limiter Based on Coated Conductors for Medium Voltage Application." *IEEE Transactions on Applied Superconductivity* **21**, 1209–1212 (2011). http://dx.doi.org/10.1109/TASC.2010.2100799. (Cited on pages 112 and 113)

[FFG10]   Farinon, S., Fabbricatore, P. and Gömöry, F. "Critical state and magnetization loss in multifilamentary superconducting wire solved through the commercial finite element code ANSYS." *Superconductor Science and Technology* **23**, 115 004 (2010). http://dx.doi.org/10.1088/0953-2048/23/11/115004. (Cited on pages 4, 83 and 87)

[FGLL93]   Feigel'Man, M.V., Geshkenbein, V.B., Larkin, A.I. and Levit, S. "Hall tunneling of vortices in high-temperature superconductors." *Journal of Experimental and Theoretical Physics Letters C/C of Pisma v Zhurnal Eksperimentalnoi i Teoreticheskoi Fiziki* **57**, 711–716 (1993). (Cited on page 17)

[FOS+11]   Fukui, S., Ogawa, J., Sato, T., Tsukamoto, O., Kashima, N. and Nagaya, S. "Study of 10 MW-Class Wind Turbine Synchronous Generators With HTS Field Windings." *IEEE Transactions on Applied Superconductivity* **21**, 1151–1154 (2011). http://dx.doi.org/10.1109/TASC.2010.2090115. (Cited on page 1)

[GAF+02]   Groves, J.R., Arendt, P.N., Foltyn, S.R., Jia, Q., Holesinger, T.G., Kung, H., DePaula, R.F., Dowden, P.C., Peterson, E.J., Stan, L. and Emmert, L.a. "Recent progress in continuously processed IBAD MgO template meters for HTS applications." *Physica C: Superconductivity* **382**, 43–47 (2002). http://dx.doi.org/10.1016/S0921-4534(02)01194-2. (Cited on page 2)

[GAS06]   Grilli, F., Ashworth, S.P. and Stavrev, S. "Magnetization AC losses of stacks of YBCO coated conductors." *Physica C: Superconductivity* **434**, 185–190 (2006). http://dx.doi.org/10.1016/j.physc.2005.12.064. (Cited on page 106)

[GBF+06]   Gianni, L., Bindi, M., Fontana, F., Ginocchio, S., Martini, L., Perini, E. and Zannella, S. "Transport AC Losses in YBCO Coated Conductors." *IEEE Transactions on Applied Superconductivity* **16**, 147–149 (2006). http://dx.doi.org/10.1109/TASC.2006.870819. (Cited on page 29)

[Gen02]   Genenko, Y.A. "Overcritical states of a superconductor strip in all-superconducting environments." *Physical Review B* **66** (2002). http://dx.doi.org/10.1103/PhysRevB.66.184520. (Cited on pages 30, 100 and 130)

[GF78]   Gray, K.E. and Fowler, D.E. "A superconducting fault-current limiter."
             Journal of Applied Physics **49**, 2546 (1978). http://dx.doi.org/10.
             1063/1.325061. (Cited on page 1)

[GFH⁺08]   Goldacker, W., Frank, A., Heller, R., Ringsdorf, B., Schlachter, S.I.,
             Kling, A., Schmidt, C., Balachandran, U., Amm, K., Evans, D., Gregory,
             E., Lee, P., Osofsky, M., Pamidi, S., Park, C., Wu, J. and Sumption, M.D.
             "Critical Currents in Roebel Assembled Coated Conductors (RACC)."
             AIP Conference Proceedings **986**, 461–470 (2008). http://dx.doi.org/
             10.1063/1.2900383. (Cited on page 2)

[GH05]   Gu, C. and Han, Z. "Calculation of AC Losses in HTS Tape With FEA
             Program ANSYS." *IEEE Transactions on Applied Superconductivity* **15**,
             2859–2862 (2005). http://dx.doi.org/10.1109/TASC.2005.848248.
             (Cited on page 83)

[GK06]   Gömöry, F. and Klinčok, B. "Self-field critical current of a conductor
             with an elliptical cross-section." *Superconductor Science and Technology*
             19, 732–737 (2006). http://dx.doi.org/10.1088/0953-2048/19/8/
             007. (Cited on pages 35 and 42)

[GL50]   Ginzburg, V.L. and Landau, L.D. "On the Theory of Superconductivity."
             Pisma v Zhurnal Eksperimentalnoi i Teoreticheskoi Fiziki **20**, 1064–1082
             (1950). (Cited on pages 6 and 7)

[GLG⁺07]   Gutiérrez, J., Llordés, A., Gázquez, J., Gibert, M., Romà, N., Ricart,
             S., Pomar, A., Sandiumenge, F., Mestres, N., Puig, T. and Obradors,
             X. "Strong isotropic flux pinning in solution-derived $YBa_2Cu_2O_{7-x}$
             nanocomposite superconductor films." *Nature materials* **6**, 367–73
             (2007). http://dx.doi.org/10.1038/nmat1893. (Cited on page 35)

[GMB64]   Gor'kov, L.P. and Melik-Barkhudarov, T.K. "Microscopic Derivation of
             the Ginzburg-Landau Equations for an Anisotropic Superconductor."
             *Journal of experimental and theoretical physics of the Academy of Sci-
             ences of the USSR* **18**, 1493–1498 (1964). (Cited on page 7)

[GNB⁺96]   Goyal, A., Norton, D.P., Budai, J.D., Paranthaman, M., Specht, E.D.,
             Kroeger, D.M., Christen, D.K., He, Q., Saffian, B., List, F.a., Lee, D.F.,
             Martin, P.M., Klabunde, C.E., Hartfield, E. and Sikka, V.K. "High
             critical current density superconducting tapes by epitaxial deposition
             of $YBa_2Cu_3O_x$ thick films on biaxially textured metals." *Applied Physics*

Letters **69**, 1795 (1996). http://dx.doi.org/10.1063/1.117489. (Cited on pages 2, 29 and 85)

[GNK+97]   Goyal, A., Norton, D.P., Kroeger, D.M., Christen, D.K., Paranthaman, M., Specht, E.D., Budai, J.D., He, Q., Saffian, B., List, F.a., Lee, D.F., Hatfield, E., Martin, P.M., Klabunde, C.E., Mathis, J. and Park, C. "Conductors with controlled grain boundaries: An approach to the next generation, high temperature superconducting wire." *Journal of Materials Research* **12**, 2924–2940 (1997). http://dx.doi.org/10.1557/JMR.1997.0387. (Cited on pages 2, 29 and 85)

[GNK+06]   Goldacker, W., Nast, R., Kotzyba, G., Schlachter, S.I., Frank, A., Ringsdorf, B., Schmidt, C. and Komarek, P. "High current DyBCO-ROEBEL Assembled Coated Conductor (RACC)." *Journal of Physics: Conference Series* **43**, 901–904 (2006). http://dx.doi.org/10.1088/1742-6596/43/1/220. (Cited on page 135)

[Gor59]   Gor'kov, L.P. "Microscopic derivation of the Ginzburg-Landau equations in the theory of superconductivity." *Journal of experimental and theoretical physics of the Academy of Sciences of the USSR* **9**, 1364–1367 (1959). (Cited on page 7)

[GP10]   Grilli, F. and Pardo, E. "Simulation of ac loss in Roebel coated conductor cables." *Superconductor Science and Technology* **23**, 115 018 (2010). http://dx.doi.org/10.1088/0953-2048/23/11/115018. (Cited on page 155)

[GPS+14]   Grilli, F., Pardo, E., Stenvall, A., Nguyen, D.N., Yuan, W. and Gomory, F. "Computation of Losses in HTS Under the Action of Varying Magnetic Fields and Currents." *IEEE Transactions on Applied Superconductivity* **24**, 1–33 (2014). http://dx.doi.org/10.1109/TASC.2013.2259827. (Cited on page 3)

[GRK11]   Genenko, Y.A., Rauh, H. and Krüger, P.A.C. "Finite-element simulations of hysteretic ac losses in a bilayer superconductor/ferromagnet heterostructure subject to an oscillating transverse magnetic field." *Applied Physics Letters* **98**, 152 508 (2011). http://dx.doi.org/10.1063/1.3560461. (Cited on pages 78 and 91)

[GRKN09]   Genenko, Y.A., Rauh, H., Krüger, P.A.C. and Narayanan, N. "Finite-element simulations of overcritical states of a magnetically shielded superconductor strip." *Superconductor Science and Technology* **22**, 055 001 (2009). http://dx.doi.org/10.1088/0953-2048/22/5/055001. (Cited on pages 4, 30, 87, 100 and 130)

[GSB$^+$11]   Glasson, N., Staines, M., Buckley, R., Pannu, M. and Kalsi, S. "Development of a 1 MVA 3-Phase Superconducting Transformer Using YBCO Roebel Cable." *IEEE Transactions on Applied Superconductivity* **21**, 1393–1396 (2011). http://dx.doi.org/10.1109/TASC.2010.2087738. (Cited on page 1)

[GSF00]   Genenko, Y.A., Snezhko, A. and Freyhardt, H.C. "Overcritical states of a superconductor strip in a magnetic environment." *Physical Review B* **62**, 3453–3472 (2000). http://dx.doi.org/10.1103/PhysRevB.62.3453. (Cited on pages 30, 100 and 130)

[GUF99]   Genenko, Y.A., Usoskin, A. and Freyhardt, H.C. "Large Predicted Self-Field Critical Current Enhancements In Superconducting Strips Using Magnetic Screens." *Physical Review Letters* **83**, 3045–3048 (1999). http://dx.doi.org/10.1103/PhysRevLett.83.3045. (Cited on page 30)

[GVPv09]   Gömöry, F., Vojenčiak, M., Pardo, E. and Šouc, J. "Magnetic flux penetration and AC loss in a composite superconducting wire with ferromagnetic parts." *Superconductor Science and Technology* **22**, 034 017 (2009). http://dx.doi.org/10.1088/0953-2048/22/3/034017. (Cited on page 30)

[GvS$^+$08]   Gömöry, F., Šouc, J., Seiler, E., Vojenčiak, M. and Granados, X. "Modification of critical current in HTSC tape conductors by a ferromagnetic layer." *Journal of Physics: Conference Series* **97**, 012 096/1–6 (2008). http://dx.doi.org/10.1088/1742-6596/97/1/012096. (Cited on page 127)

[GvV$^+$07]   Gömöry, F., Šouc, J., Vojenčiak, M., Alamgir, A.K.M., Han, Z. and Gu, C. "Reduction of ac transport and magnetization loss of a high-T_c superconducting tape by placing soft ferromagnetic materials at the edges." *Applied Physics Letters* **90**, 092 506 (2007). http://dx.doi.org/10.1063/1.2710753. (Cited on pages 27, 91 and 127)

[Haz13]   Hazelton, D.W. "2G HTS Conductor Development at SuperPower for Magnet Applications." (2013). (Cited on page 3)

[HC10]   Hong, Z. and Coombs, T.A. "Numerical Modelling of AC Loss in Coated Conductors by Finite Element Software Using H Formulation." *Journal of Superconductivity and Novel Magnetism* **23**, 1551–1562 (2010). http://dx.doi.org/10.1007/s10948-010-0812-y. (Cited on page 42)

[HCC06]   Hong, Z., Campbell, A.M. and Coombs, T.A. "Numerical solution of critical state in superconductivity by finite element software." *Superconductor Science and Technology* **19**, 1246–1252 (2006). http://dx.doi.org/10.1088/0953-2048/19/12/004. (Cited on pages 4 and 42)

[HEB+12]   Hobl, A., Elschner, S., Bock, J., Kramer, S., Janke, C. and Schramm, J. "Superconducting fault current limiters: a new tool for the "grid of the future"." In "CIRED 2012 Workshop: Integration of Renewables into the Distribution Grid," May, pp. 296–296. IET (2012). ISBN 978-1-84919-628-4. http://dx.doi.org/10.1049/cp.2012.0852. (Cited on page 1)

[HHJ+04]   Hassenzahl, W.V., Hazelton, D.W., Johnson, B.K., Komarek, P., Noe, M. and Reis, C.T. "Electric power applications of superconductivity." *Proceedings of the IEEE* **92**, 1655–1674 (2004). http://dx.doi.org/10.1109/JPROC.2004.833674. (Cited on pages 2 and 134)

[HKC+07]   Hänisch, J., Kozlova, N., Cai, C., Nenkov, K., Fuchs, G. and Holzapfel, B. "Determination of the irreversibility field of YBCO thin films from pulsed high-field measurements." *Superconductor Science and Technology* **20**, 228–231 (2007). http://dx.doi.org/10.1088/0953-2048/20/3/019. (Cited on page 9)

[HM04]   Hull, J. and Murakami, M. "Applications of bulk high-temperature Superconductors." *Proceedings of the IEEE* **92**, 1705–1718 (2004). http://dx.doi.org/10.1109/JPROC.2004.833796. (Cited on page 2)

[Hos52]   Hoselitz, K. *Ferromagnetic Properties of Metals and Alloys*. Clarendon Press, Oxford (1952). (Cited on page 30)

[HSM+91]   Hashizume, H., Sugiura, T., Miya, K., Ando, Y., Akita, S., Torii, S., Kubota, Y. and Ogasawara, T. "Numerical analysis of a.c. losses in superconductors." *Cryogenics* **31**, 601–606 (1991). http://dx.doi.org/10.1016/0011-2275(91)90057-4. (Cited on pages 4 and 42)

[HTT89]   Heine, K., Tenbrink, J. and Thöner, M. "High-field critical current densities in Bi2Sr2Ca1Cu2O8+x/Ag wires." *Applied Physics Letters* **55**, 2441 (1989). http://dx.doi.org/10.1063/1.102295. (Cited on page 2)

[Hua89]   Huang, C.L. "Temperature dependence of contact resistance in noble metal/high-Tc superconductor systems." *Journal of Vacuum Science & Technology A: Vacuum, Surfaces, and Films* **7**, 3371 (1989). http://dx.doi.org/10.1116/1.576152. (Cited on page 141)

[HYT⁺00]   Hashimoto, H., Yamada, T., Tani, K., Honjo, S., Sato, Y. and Ishii, H. "Finite element analysis of AC losses in double helix superconducting cables." *IEEE Transactions on Magnetics* **36**, 1205–1208 (2000). http://dx.doi.org/10.1109/20.877656. (Cited on pages 4 and 42)

[IHO⁺09]   Iwakuma, M., Hayashi, H., Okamoto, H., Tomioka, A., Konno, M., Saito, T., Iijima, Y., Suzuki, Y., Yoshida, S., Yamada, Y., Izumi, T. and Shiohara, Y. "Development of REBCO superconducting power transformers in Japan." *Physica C: Superconductivity* **469**, 1726–1732 (2009). http://dx.doi.org/10.1016/j.physc.2009.05.246. (Cited on page 1)

[ITTT87]   Iye, Y., Tamegai, T., Takeya, H. and Takei, H. "The Anisotropic Upper Critical Field of Single Crystal $YBa_2Cu_2O_x$." *Japanese Journal of Applied Physics* **26**, L1057–L1059 (1987). http://dx.doi.org/10.1143/JJAP.26.L1057. (Cited on page 35)

[JBH05]   Jooss, C., Brinkmeier, E. and Heese, H. "Combined experimental and theoretical study of field and current conditioning in magnetically shielded superconducting films." *Physical Review B* **72**, 1–12 (2005). http://dx.doi.org/10.1103/PhysRevB.72.144516. (Cited on pages 30, 100 and 130)

[JGK⁺02]   Juengst, K.P., Gehring, R., Kudymow, A., Kuperman, G. and Suess, E. "25 MW SMES-based power modulator." *IEEE Transactions on Appiled Superconductivity* **12**, 758–761 (2002). http://dx.doi.org/10.1109/TASC.2002.1018512. (Cited on page 1)

[JJF02]   Jarzina, H., Jooss, C. and Freyhardt, H.C. "Overcritical Meissner current densities in $YBa_2Cu_2O_7$ films in soft magnetic environments." *Journal of Applied Physics* **91**, 3775 (2002). http://dx.doi.org/10.1063/1.1454221. (Cited on pages 30, 100 and 130)

[JLL03]   Jewell, M.C., Lee, P.J. and Larbalestier, D.C. "The influence of Nb_3Sn strand geometry on filament breakage under bend strain as revealed by metallography." *Superconductor Science and Technology* **16**, 1005–1011 (2003). http://dx.doi.org/10.1088/0953-2048/16/9/308. (Cited on page 22)

[JMRB80] Jérome, D., Mazaud, A., Ribault, M. and Bechgaard, K. "Superconductivity in a synthetic organic conductor $(TMTSF)_2PF_6$." *Journal de Physique Lettres* **41**, 95–98 (1980). http://dx.doi.org/10.1051/jphyslet:0198000410409500. (Cited on page 20)

[Kaw01] Kawasaki, K. "Theoretical expressions for AC losses of superconducting coils in external magnetic field and transport current with phase difference." *Physica C: Superconductivity* **357-360**, 1205–1208 (2001). http://dx.doi.org/10.1016/S0921-4534(01)00486-5. (Cited on page 2)

[KC07] Kasap, S. and Capper, P. *Springer Handbook of Electronic and Photonic Materials.* Springer US, Boston, MA, 1st ed. (2007). ISBN 978-0-387-26059-4. http://dx.doi.org/10.1007/978-0-387-29185-7. (Cited on page 3)

[KGF11] Krüger, P.A.C., Grilli, F. and Farinon, S. "Compliance of numerical formulations for describing superconductor/ferromagnet heterostructures." *Physica C: Superconductivity* **471**, 1083–1085 (2011). http://dx.doi.org/10.1016/j.physc.2011.05.129. (Cited on page 87)

[KHF$^+$06] Kamijo, H., Hata, H., Fujimoto, H., Inoue, a., Nagashima, K., Ikeda, K., Yamada, H., Sanuki, Y., Tomioka, a., Uwamori, K., Yoshida, S., Iwakuma, M. and Funaki, K. "Fabrication of Superconducting Traction Transformer for Railway Rolling Stock." *Journal of Physics: Conference Series* **43**, 841–844 (2006). http://dx.doi.org/10.1088/1742-6596/43/1/205. (Cited on page 1)

[KHH$^+$02] Kimura, H., Honda, K., Hayashi, H., Tsutsumi, K., Iwakuma, M., Funaki, K., Bohno, T., Tomioka, A., Yagi, Y., Maruyama, H. and Ohashi, K. "Test results of a HTS power transformer connected to a power grid." *Physica C: Superconductivity* **372-376**, 1694–1697 (2002). http://dx.doi.org/10.1016/S0921-4534(02)01103-6. (Cited on page 1)

[KHS62] Kim, Y.B., Hempstead, C.F. and Strnad, A.R. "Critical persistent currents in hard superconductors." *Physical Review Letters* **9**, 306–309 (1962). (Cited on page 35)

[KHY+03]   Kajikawa, K., Hayashi, T., Yoshida, R., Iwakuma, M. and Funaki, K. "Numerical evaluation of AC losses in HTS wires with 2D FEM formulated by self magnetic field." *IEEE Transactions on Applied Superconductivity* **13**, 3630–3633 (2003). http://dx.doi.org/10.1109/TASC.2003.812415. (Cited on pages 4, 42 and 44)

[Kit96]   Kittel, C. *Introduction to Solid State Physics*. John Wiley & Sons, Inc., New York, 7th ed. (1996). ISBN 0-471-11181-3. (Cited on page 9)

[KJL+05]   Kim, D.W., Jang, H.M., Lee, C.h., Kim, J.H., Ha, C.W., Kwon, Y.H. and Cho, J.W. "Development of the 22.9-kV Class HTS Power Cable in LG Cable." *IEEE Transactions on Appiled Superconductivity* **15**, 1723–1726 (2005). http://dx.doi.org/10.1109/TASC.2005.849266. (Cited on pages 2 and 134)

[KKK+11]   Kojima, H., Kotari, M., Kito, T., Hayakawa, N., Hanai, M. and Okubo, H. "Current Limiting and Recovery Characteristics of 2 MVA Class Superconducting Fault Current Limiting Transformer (SFCLT)." *IEEE Transactions on Applied Superconductivity* **21**, 1401–1404 (2011). http://dx.doi.org/10.1109/TASC.2010.2089413. (Cited on page 1)

[KKL+06]   Kim, W.S., Kwak, S.Y., Lee, J.K., Choi, K.D., Jung, H.K., Seong, K.C. and Hahn, S. "Design of HTS Magnets for a 600 kJ SMES." *IEEE Transactions on Applied Superconductivity* **16**, 620–623 (2006). http://dx.doi.org/10.1109/TASC.2005.864244. (Cited on page 1)

[KKT+99]   Krüger Olsen, S., Kühle, A., Traeholt, C., Rasmussen, C., Tonnesen, O., Däumling, M., Rasmussen, C.N. and Willen, D.W.A. "Alternating current losses of a 10 metre long low loss superconducting cable conductor determined from phase sensitive measurements." *Superconductor Science and Technology* **12**, 360–365 (1999). http://dx.doi.org/10.1088/0953-2048/12/6/306. (Cited on pages 2 and 134)

[KLL+09]   Kwak, S., Lee, S., Lee, S., Kim, W.s., Lee, J.k., Park, C., Bae, J., Song, J.b., Lee, H., Choi, K., Seong, K., Jung, H. and Hahn, S.y. "Design of HTS Magnets for a 2.5 MJ SMES." *IEEE Transactions on Applied Superconductivity* **19**, 1985–1988 (2009). http://dx.doi.org/10.1109/TASC.2009.2018754. (Cited on page 1)

[KNN+09]   Katagiri, T., Nakabayashi, H., Nijo, Y., Tamada, T., Noda, T., Hirano, N., Nagata, T., Nagaya, S., Yamane, M., Ishii, Y. and Nitta, T. "Field Test

Result of 10MVA/20MJ SMES for Load Fluctuation Compensation." *IEEE Transactions on Applied Superconductivity* **19**, 1993–1998 (2009). http://dx.doi.org/10.1109/TASC.2009.2018479. (Cited on page 1)

[KSC+06] Kim, H., Seong, K., Cho, J., Bae, J., Sim, K., Kim, S., Lee, E., Ryu, K. and Kim, S. "3 MJ/750 kVA SMES System for Improving Power Quality." *IEEE Transactions on Applied Superconductivity* **16**, 574–577 (2006). http://dx.doi.org/10.1109/TASC.2006.871329. (Cited on page 1)

[KSC+12] Kraemer, H.P., Schmidt, W., Cai, H., Gamble, B., Madura, D., MacDonald, T., McNamara, J., Romanosky, W., Snitchler, G., Lallouet, N., Schmidt, F. and Ahmed, S. "Superconducting Fault Current Limiter for Transmission Voltage." *Physics Procedia* **36**, 921–926 (2012). http://dx.doi.org/10.1016/j.phpro.2012.06.230. (Cited on page 1)

[KTA+96] Kasahara, H., Torii, S., Akita, S., Kubota, Y., Yasohama, K., Kobayashi, H., Ogasawara, T. and Kumano, T. "Development of an Nb_3Sn AC coil with react & wind method." *IEEE Transactions on Magnetics* **32**, 2874–2878 (1996). http://dx.doi.org/10.1109/20.511475. (Cited on page 22)

[KTK+01] Kajikawa, K., Takenaka, A., Kawasaki, K., Iwakuma, M. and Funaki, K. "Numerical simulation for AC losses of HTS tapes in combined alternating transport current and external AC magnetic field with phase shift." *IEEE Transactions on Applied Superconductivity* **11**, 2240–2243 (2001). http://dx.doi.org/10.1109/77.920305. (Cited on pages 4 and 42)

[KWN+01] Kelley, N., Wakefield, C., Nassi, M., Corsaro, P., Spreafico, S., Von Dollen, D. and Jipping, J. "Field demonstration of a 24-kV warm dielectric HTS cable." *IEEE Transactions on Appiled Superconductivity* **11**, 2461–2466 (2001). http://dx.doi.org/10.1109/77.920361. (Cited on pages 2 and 134)

[LGFP01] Larbalestier, D.C., Gurevich, A., Feldmann, D.M. and Polyanskii, A. "High-Tc superconducting materials for electric power applications." *Nature* **414**, 368–77 (2001). http://dx.doi.org/10.1038/35104654. (Cited on pages 1 and 2)

[LL35]   London, F. and London, H. "The Electromagnetic Equations of the Supraconductor." *Proceedings of the Royal Society of London* **149**, 71–88 (1935). (Cited on page 6)

[LM07]   Lewis, C. and Muller, J. "A Direct Drive Wind Turbine HTS Generator." In "2007 IEEE Power Engineering Society General Meeting," pp. 1–8. IEEE (2007). ISBN 1-4244-1296-X. http://dx.doi.org/10.1109/PES.2007.386069. (Cited on page 1)

[LMN+09]   Luongo, C.A., Masson, P.J., Nam, T., Mavris, D., Kim, H.D., Brown, G.V., Waters, M. and Hall, D. "Next Generation More-Electric Aircraft: A Potential Application for HTS Superconductors." *IEEE Transactions on Applied Superconductivity* **19**, 1055–1068 (2009). http://dx.doi.org/10.1109/TASC.2009.2019021. (Cited on page 1)

[Lon63]   London, H. "Alternating current losses in superconductors of the second kind." *Physics Letters* **6**, 162–165 (1963). http://dx.doi.org/10.1016/0031-9163(63)90527-4. (Cited on page 33)

[M99]   Müller, K.H. "AC losses in stacks and arrays of YBCO/hastelloy and monofilamentary Bi-2223/Ag tapes." *Physica C: Superconductivity* pp. 149–167 (1999). (Cited on page 106)

[Mal06]   Malozemoff, A.P. "Progress in HTS Power Applications in the United States." *Teion Kogaku (Journal of the Cryogenic Society of Japan)* **41**, 164–169 (2006). http://dx.doi.org/10.2221/jcsj.41.164. (Cited on pages 2 and 134)

[Mal12]   Malozemoff, A.P. "Second-Generation High-Temperature Superconductor Wires for the Electric Power Grid." *Annual Review of Materials Research* **42**, 373–397 (2012). http://dx.doi.org/10.1146/annurev-matsci-100511-100240. (Cited on page 1)

[Mas02]   Masuda, T. "Experimental results of a 30 m, 3-core HTSC cable." *Physica C: Superconductivity* **372-376**, 1555–1559 (2002). http://dx.doi.org/10.1016/S0921-4534(02)01062-6. (Cited on pages 2 and 134)

[Maw96]   Mawatari, Y. "Critical state of periodically arranged superconducting-strip lines in perpendicular fields." *Physical Review B* **54**, 13 215–13 221 (1996). http://dx.doi.org/10.1103/PhysRevB.54.13215. (Cited on pages 2 and 106)

[Maw08]   Mawatari, Y. "Magnetic field distributions around superconducting strips on ferromagnetic substrates." *Physical Review B* **77**, 1–8 (2008). http://dx.doi.org/10.1103/PhysRevB.77.104505. (Cited on pages 2, 3 and 41)

[MBJC13]   Melhem, Z., Ball, S., Jokinen, A. and Chappell, S.P.G. "Electromagnetic Characteristics of Bi-2212 Test Coils at High Field During Quench." *IEEE Transactions on Applied Superconductivity* **23**, 4603 104–4603 104 (2013). http://dx.doi.org/10.1109/TASC.2013.2238281. (Cited on page 2)

[MBSL07]   Masson, P.J., Brown, G.V., Soban, D.S. and Luongo, C.A. "HTS machines as enabling technology for all-electric airborne vehicles." *Superconductor Science and Technology* **20**, 748–756 (2007). http://dx.doi.org/10.1088/0953-2048/20/8/005. (Cited on page 1)

[MC01]   Mawatari, Y. and Clem, J.R. "Magnetic-Flux Penetration and Critical Currents in Superconducting Strips with Slits." *Physical Review Letters* **86**, 2870–2873 (2001). http://dx.doi.org/10.1103/PhysRevLett.86. 2870. (Cited on page 2)

[MCT⁺09]   Masson, P.J., Choi, T.P., Tixador, P., Waters, M., Hall, D., Luongo, C.A. and Mavris, D.N. "Superconducting Ducted Fan Design for Reduced Emissions Aeropropulsion." *IEEE Transactions on Applied Superconductivity* **19**, 1662–1668 (2009). http://dx.doi.org/10.1109/TASC.2009. 2018156. (Cited on page 1)

[MFL⁺09]   Maguire, J., Folts, D., Lindsay, D., Knoll, D., Bratt, S., Wolff, Z. and Kurtz, S. "Development and Demonstration of a Fault Current Limiting HTS Cable to be Installed in the Con Edison Grid." *IEEE Transactions on Applied Superconductivity* **19**, 1740–1743 (2009). http://dx.doi. org/10.1109/TASC.2009.2018313. (Cited on pages 2 and 134)

[MFR⁺08]   Malozemoff, A.P., Fleshler, S., Rupich, M., Thieme, C.L.H., Li, X., Zhang, W., Otto, A., Maguire, J., Folts, D., Yuan, J., Kraemer, H.P., Schmidt, W., Wohlfart, M. and Neumueller, H.W. "Progress in high temperature superconductor coated conductors and their applications." *Superconductor Science and Technology* **21**, 034 005 (2008). http://dx.doi.org/10.1088/0953-2048/21/3/034005. (Cited on pages 2 and 29)

[MGC⁺06] Majoros, M., Glowacki, B.A., Campbell, A.M., Levin, G.A. and Barnes, P.N. "Transport AC losses in striated YBCO coated conductors." *Journal of Physics: Conference Series* **43**, 564–567 (2006). http://dx.doi.org/10.1088/1742-6596/43/1/139. (Cited on page 29)

[MH12]   Matias, V. and Hammond, R.H. "YBCO Superconductor Wire based on IBAD-Textured Templates and RCE of YBCO: Process Economics." *Physics Procedia* **36**, 1440–1444 (2012). http://dx.doi.org/10.1016/j.phpro.2012.06.239. (Cited on page 3)

[MJBM88] May, P., Jedamzik, D., Boyle, W. and Miller, P. "Production of superconducting thick films by a spin-on process." *Superconductor Science and Technology* **1**, 1–4 (1988). http://dx.doi.org/10.1088/0953-2048/1/1/001. (Cited on page 2)

[MKH⁺09] Mimura, T., Kitoh, Y., Honjo, S., Ohya, M., Yumura, H. and Masuda, T. "Outline of a new HTS cable project in Yokohama." *Physica C: Superconductivity* **469**, 1697–1701 (2009). http://dx.doi.org/10.1016/j.physc.2009.05.030. (Cited on pages 2 and 134)

[MKO⁺12] Matsumoto, S., Kiyoshi, T., Otsuka, A., Hamada, M., Maeda, H., Yanagisawa, Y., Nakagome, H. and Suematsu, H. "Generation of 24 T at 4.2 K using a layer-wound GdBCO insert coil with Nb_3Sn and NbTi external magnetic field coils." *Superconductor Science and Technology* **25**, 025 017 (2012). http://dx.doi.org/10.1088/0953-2048/25/2/025017. (Cited on page 2)

[MKY⁺02] Masuda, T., Kato, T., Yumura, H., Watanabe, M., Ashibe, Y. and Ohkura, K. "Verification tests of a 66 kV HTSC cable system for practical use (first cooling tests)." *Physica C: Superconductivity* **381**, 1174–1180 (2002). (Cited on pages 2 and 134)

[MLW⁺12] Markiewicz, W.D., Larbalestier, D.C., Weijers, H.W., Voran, A.J., Pickard, K.W., Sheppard, W.R., Jaroszynski, J., Walsh, R.P., Gavrilin, A.V. and Noyes, P.D. "Design of a Superconducting 32 T Magnet With REBCO High Field Coils." *IEEE Transactions on Applied Superconductivity* **22**, 4300 704–4300 704 (2012). http://dx.doi.org/10.1109/TASC.2011.2174952. (Cited on page 2)

[MO33]   Meissner, W. and Ochsenfeld, R. "Ein neuer Effekt bei Eintritt der Supraleitfähigkeit." *Die Naturwissenschaften* **21**, 787–788 (1933). http://dx.doi.org/10.1007/BF01504252. (Cited on page 5)

[MSB+07] Maguire, J.F., Schmidt, F., Bratt, S., Welsh, T.E., Yuan, J., Allais, a. and Hamber, F. "Development and Demonstration of a HTS Power Cable to Operate in the Long Island Power Authority Transmission Grid." *IEEE Transactions on Applied Superconductivity* **17**, 2034–2037 (2007). http://dx.doi.org/10.1109/TASC.2007.898359. (Cited on pages 2 and 134)

[MSBW09] Maguire, J., Schmidt, F., Bratt, S. and Welsh, T. "Installation and Testing Results of Long Island Transmission Level HTS Cable." *IEEE Transactions on Applied Superconductivity* **19**, 1692–1697 (2009). http://dx.doi.org/10.1109/TASC.2009.2018221. (Cited on pages 2 and 134)

[MSU+13] Miller, J., Santosusso, D., Uva, M., Woods, K. and Fitzpatrick, B. "Naval Superconducting Integrated Power System." In "Proceedings of the American Society of Naval Engineers: Intelligent Ships Symposium X," vol. Intelligen, pp. 1–8 (2013). (Cited on page 1)

[MU97] Müller, P. and Ustinov, A.V. (Editors). *The Physics of Superconductors*. Springer-Verlag, Berlin/Heidelberg, 1st ed. (1997). ISBN 3-540-61243-2. (Cited on pages 6, 7, 10, 13 and 18)

[Mur94] Mur, G. "Edge elements, their advantages and their disadvantages." *IEEE Transactions on Magnetics* **30**, 3552–3557 (1994). http://dx.doi.org/10.1109/20.312706. (Cited on pages 59 and 60)

[MYU+08] Miyagi, D., Yunoki, Y., Umabuchi, M., Takahashi, N. and Tsukamoto, O. "Measurement of magnetic properties of Ni-alloy substrate of HTS coated conductor in LN2." *Physica C: Superconductivity* **468**, 1743–1746 (2008). http://dx.doi.org/10.1016/j.physc.2008.05.196. (Cited on page 29)

[MYV+07] Majoros, M., Ye, L., Velichko, A.V., Coombs, T.A., Sumption, M.D. and Collings, E.W. "Transport AC losses in YBCO coated conductors." *Superconductor Science and Technology* **20**, S299–S304 (2007). http://dx.doi.org/10.1088/0953-2048/20/9/S27. (Cited on page 29)

[NAW09] Nguyen, D.N., Ashworth, S.P. and Willis, J.O. "Experimental and finite-element method studies of the effects of ferromagnetic substrate on the total ac loss in a rolling-assisted biaxially textured substrate $YBa_2Cu_3O_7$ tape exposed to a parallel ac magnet." *Journal of Applied*

Physics **106**, 093 913 (2009). http://dx.doi.org/10.1063/1.3255998. (Cited on pages 4 and 42)

[NAW⁺10]   Nguyen, D.N., Ashworth, S.P., Willis, J.O., Sirois, F. and Grilli, F. "A new finite-element method simulation model for computing AC loss in roll assisted biaxially textured substrate YBCO tapes." *Superconductor Science and Technology* **23**, 025 001 (2010). http://dx.doi.org/10. 1088/0953-2048/23/2/025001. (Cited on page 51)

[NEM⁺13]   Nishijima, S., Eckroad, S., Marian, A., Choi, K., Kim, W.S., Terai, M., Deng, Z., Zheng, J., Wang, J., Umemoto, K., Du, J., Febvre, P., Keenan, S., Mukhanov, O., Cooley, L.D., Foley, C.P., Hassenzahl, W.V. and Izumi, M. "Superconductivity and the environment: a Roadmap." *Superconductor Science and Technology* **26**, 113 001 (2013). http://dx.doi.org/10.1088/0953-2048/26/11/113001. (Cited on page 1)

[NFK⁺10]   Nick, W., Frank, M., Kummeth, P., Rabbers, J.J., Wilke, M. and Schleicher, K. "Development and construction of an HTS rotor for ship propulsion application." *Journal of Physics: Conference Series* **234**, 032 040 (2010). http://dx.doi.org/10.1088/1742-6596/234/3/032040. (Cited on page 1)

[NHK⁺04]   Nagaya, S., Hirano, N., Kondo, M., Tanaka, T., Nakabayashi, H., Shikimachi, K., Hanai, S., Inagaki, J., Ioka, S. and Kawashima, S. "Development and Performance Results of 5 MVA SMES for Bridging Instantaneous Voltage Dips." *IEEE Transactions on Appiled Superconductivity* **14**, 699–704 (2004). http://dx.doi.org/10.1109/TASC.2004.830076. (Cited on page 1)

[NHS⁺04]   Nagaya, S., Hirano, N., Shikimachi, K., Hanai, S., Inagaki, J., Maruyama, K., Ioka, S., Ono, M., Ohsemochi, K. and Kurusu, T. "Development of MJ-Class HTS SMES for Bridging Instantaneous Voltage Dips." *IEEE Transactions on Appiled Superconductivity* **14**, 770–773 (2004). http://dx.doi.org/10.1109/TASC.2004.830105. (Cited on page 1)

[NNM⁺01]   Nagamatsu, J., Nakagawa, N., Muranaka, T., Zenitani, Y. and Akimitsu, J. "Superconductivity at 39 K in magnesium diboride." *Nature* **410**, 63–4 (2001). http://dx.doi.org/10.1038/35065039. (Cited on pages 1 and 22)

[Nor70]   Norris, W.T. "Calculation of hysteresis losses in hard superconductors carrying ac: isolated conductors and edges of thin sheets." *Journal of Physics D: Applied Physics* **3**, 489 (1970). (Cited on pages 2, 3, 37 and 82)

[NS07]   Noe, M. and Steurer, M. "High-temperature superconductor fault current limiters: concepts, applications, and development status." *Superconductor Science and Technology* **20**, R15–R29 (2007). http://dx.doi.org/10.1088/0953-2048/20/3/R01. (Cited on page 1)

[NSF12]   Naito, T., Sasaki, T. and Fujishiro, H. "Trapped magnetic field and vortex pinning properties of MgB 2 superconducting bulk fabricated by a capsule method." *Superconductor Science and Technology* **25**, 095 012 (2012). http://dx.doi.org/10.1088/0953-2048/25/9/095012. (Cited on page 2)

[NTFA89]   Nakata, T., Takahashi, N., Fujiwara, K. and Ahagon, A. "3-D non-linear eddy current analysis using the time-periodic finite element method." *IEEE Transactions on Magnetics* **25**, 4150–4152 (1989). http://dx.doi.org/10.1109/20.42551. (Cited on pages 4 and 42)

[OCA$^+$06] Ottonello, L., Canepa, G., Albertelli, P., Picco, E., Florio, A., Masciarelli, G., Rossi, S., Martini, L., Pincella, C., Mariscotti, A., Torello, E., Martinolli, A. and Mariani, M. "The Largest Italian SMES." *IEEE Transactions on Applied Superconductivity* **16**, 602–607 (2006). http://dx.doi.org/10.1109/TASC.2005.869677. (Cited on page 1)

[Onn12]   Onnes, H.K. "Further experiments with Liquid Helium G. On the electrical resistance of Pure Metals etc. VI. On the Sudden Change in the Rate at which the Resistance of Mercury Disappears." *KNAW Proceedings 14 II. Amsterdam Huygens Institute - Royal Netherlands Academy of Arts and Sciences* pp. 818–821 (1912). (Cited on page 5)

[PMC$^+$03] Pecher, R., McCulloch, M.D., Chapman, S.J., Prigozhin, L. and Elliott, C.M. "3D-modelling of bulk Type-II superconductors using unconstrained H-formulation." In "Proc. 6th European Conf. on Applied Superconductivity," (2003). (Cited on pages 4 and 42)

[Pre11]   Presse- und Informationsamt der Bundesregierung. "Energiewende - die einzelnen Maßnahmen im Überblick." (2011). (Cited on page 1)

[Pri97]   Prigozhin, L. "Analysis of critical-state problems in Type-II superconductivity." *IEEE Transactions on Applied Superconductivity* **7**, 3866–

3873 (1997). http://dx.doi.org/10.1109/77.659440. (Cited on pages 4 and 42)

[PS11] Prigozhin, L. and Sokolovsky, V. "Computing AC losses in stacks of high-temperature superconducting tapes." *Superconductor Science and Technology* **24**, 075 012 (2011). http://dx.doi.org/10.1088/0953-2048/24/7/075012. (Cited on page 106)

[PVGv11] Pardo, E., Vojenčiak, M., Gömöry, F. and Šouc, J. "Low-magnetic-field dependence and anisotropy of the critical current density in coated conductors." *Superconductor Science and Technology* **24**, 065 007 (2011). http://dx.doi.org/10.1088/0953-2048/24/6/065007. (Cited on pages 35 and 36)

[RLR⁺98] Ries, G., Leghissa, M., Rieger, J., Wiezorek, J. and Oomen, M. "High-Tc superconductors and AC loss in electrotechnical devices." *Physica C: Superconductivity* **310**, 283–290 (1998). http://dx.doi.org/10.1016/S0921-4534(98)00477-8. (Cited on page 3)

[Row95] Rowley, A. "Superconducting fault current limiters." In "IEE Colloquium on High Tc Superconducting Materials as 'Magnets'," vol. 1995, pp. 10–10. IEE (1995). http://dx.doi.org/10.1049/ic:19951527. (Cited on page 1)

[Sch00] Schmidt, C. "Calorimetric ac-loss measurement of high-T_c-tapes at 77 K, a new measuring technique." *Cryogenics* **40**, 137–143 (2000). http://dx.doi.org/10.1016/S0011-2275(00)00014-X. (Cited on pages 66 and 68)

[Sch06] Schmidt, C. "Ac-loss and time constant measuring techniques of high-T_c tapes and stacks of tapes." *Physica C: Superconductivity* **445-448**, 694–700 (2006). http://dx.doi.org/10.1016/j.physc.2006.05.007. (Cited on page 106)

[SCK⁺06] Sohn, S.H., Choi, H.S., Kim, H.R., Hyun, O.B., Yim, S.W., Masuda, T., Yatsuka, K., Watanabe, M., Ryoo, H.S., Yang, H.S., Kim, D.L. and Hwang, S.D. "Field Test of 3 phase, 22.9 kV, 100 m HTS Cable System in KEPCO." *Journal of Physics: Conference Series* **43**, 885–888 (2006). http://dx.doi.org/10.1088/1742-6596/43/1/216. (Cited on pages 2 and 134)

[Seb94]   Sebestyén, I. "Numerical simulation of electromagnetic field in type-ii superconductors." *Periodica Polytechnica Series Eletrical Engineering* **38**, 57–64 (1994). (Cited on pages 4 and 42)

[SGKW11]   Snitchler, G., Gamble, B., King, C. and Winn, P. "10 MW Class Superconductor Wind Turbine Generators." *IEEE Transactions on Applied Superconductivity* **21**, 1089–1092 (2011). http://dx.doi.org/10.1109/TASC.2010.2100341. (Cited on page 1)

[SHL+07]   Sohn, S., Hwang, S., Lim, J., Yim, S., Hyun, O., Kim, H., Yatsuka, K., Masuda, T., Isojima, S., Watanabe, M., Ryoo, H., Yang, H. and Kim, D. "Verification test of 22.9kV underground HTS cable." *Physica C: Superconductivity* **463-465**, 1146–1149 (2007). http://dx.doi.org/10.1016/j.physc.2007.04.254. (Cited on pages 2 and 134)

[SIS+08]   Suenaga, M., Iwakuma, M., Sueyoshi, T., Izumi, T., Mimura, M., Takahashi, Y. and Aoki, Y. "Effects of a ferromagnetic substrate on hysteresis losses of a $YBa_2Cu_2O_7$ coated conductor in perpendicular ac applied magnetic fields." *Physica C: Superconductivity* **468**, 1714–1717 (2008). http://dx.doi.org/10.1016/j.physc.2008.05.182. (Cited on page 90)

[SM60]   Shaw, R.W. and Mapother, D.E. "Irreversibility in the Superconducting Transition of Lead." *Physical Review* **118**, 1474–1484 (1960). http://dx.doi.org/10.1103/PhysRev.118.1474. (Cited on page 16)

[SMN+12]   Stemmle, M., Merschel, F., Noe, M., Hofmann, L. and Hobl, A. "Superconducting MV cables to replace HV cables in urban area distribution grids." In "PES T&D 2012," pp. 1–5. IEEE (2012). ISBN 978-1-4673-1935-5. http://dx.doi.org/10.1109/TDC.2012.6281443. (Cited on pages 2 and 134)

[SSA+07]   Sims, R., Schock, R., Adegbululgbe, A., Fenhann, J., Konstantinaviciute, I., Moomaw, W., Nimir, H. and Schlamadinger, B. "Climate Change 2007: Working Group III: Mitigation of Climate Change." Tech. rep., IPCC Fourth Assessment Report, Cambridge, UK (2007). (Cited on page 1)

[SSS+03]   Sosolik, C.E., Stroscio, J.A., Stiles, M.D., Hudson, E.W., Blankenship, S.R., Fein, A.P. and Celotta, R.J. "Real-space imaging of structural transitions in the vortex lattice of V_3Si." *Physical Review B* **68**, 140 503 (2003). http://dx.doi.org/10.1103/PhysRevB.68.140503. (Cited on page 11)

[ST10]   Stenvall, A. and Tarhasaari, T. "An eddy current vector potential formulation for estimating hysteresis losses of superconductors with FEM." *Superconductor Science and Technology* **23**, 125013 (2010). http://dx.doi.org/10.1088/0953-2048/23/12/125013. (Cited on pages 4, 42 and 43)

[SVF$^+$10]   Sytnikov, V.E., Vysotsky, V.S., Fetisov, S.S., Nosov, A.A., Shakaryan, Y.G., Kochkin, V.I., Kiselev, A.N., Terentyev, Y.A., Patrikeev, V.M., Zubko, V.V. and Weisend, J.G. "Cryogenic and Electrical Test Results of 30 m HTS Power Cable." In "Advances in Cryogenic Engineering: Transactions of the Cryogenic Engineering Conference–CEC," vol. 55, pp. 461–468 (2010). http://dx.doi.org/10.1063/1.3422391. (Cited on pages 2 and 134)

[SZL$^+$07]   Suo, H., Zhao, Y., Liu, M., Ma, L., He, D., Zhang, Y. and Zhou, M. "Preparation of Cube Textured Ni5W/Ni9W Composite Substrate for YBCO Coated Conductors." *IEEE Transactions on Applied Superconductivity* **17**, 3420–3423 (2007). http://dx.doi.org/10.1109/TASC.2007.899594. (Cited on page 29)

[TBD$^+$07]   Tixador, P., Bellin, B., Deleglise, M., Vallier, J., Bruzek, C., Allais, A. and Saugrain, J. "Design and First Tests of a 800 kJ HTS SMES." *IEEE Transactions on Applied Superconductivity* **17**, 1967–1972 (2007). http://dx.doi.org/10.1109/TASC.2007.898520. (Cited on page 1)

[TCBM11]   Takayasu, M., Chiesa, L., Bromberg, L. and Minervini, J.V. "Cabling Method for High Current Conductors Made of HTS Tapes." *IEEE Transactions on Applied Superconductivity* **21**, 2340–2344 (2011). http://dx.doi.org/10.1109/TASC.2010.2094176. (Cited on pages 133 and 135)

[TCBM12]   Takayasu, M., Chiesa, L., Bromberg, L. and Minervini, J.V. "HTS twisted stacked-tape cable conductor." *Superconductor Science and Technology* **25**, 014011 (2012). http://dx.doi.org/10.1088/0953-2048/25/1/014011. (Cited on pages 138, 139, 140 and 141)

[Tin96]   Tinkham, M. *Introduction to Superconductivity*. McGraw-Hill, Inc., New York (1996). ISBN 0-07-064878-6. (Cited on pages 5, 7 and 18)

[TMBB10]   Takayasu, M., Minervini, J.V., Bromberg, L. and Balachandran, U. "Torsion Strain Effects on Critical Currents of HTS Superconducting

Tapes." In "AIP Conference Proceedings: Transactions of the International Cryogenic Materials Conference – ICMC," vol. 337, pp. 337–344 (2010). ISBN 0735407657. http://dx.doi.org/10.1063/1.3402320. (Cited on page 36)

[TOK⁺05] Tasaki, K., Ono, M., Kuriyama, T., Kyoto, M., Hanai, S., Takigami, H., Takano, H., Watanabe, K., Awaji, S., Nishijima, G. and Togano, K. "Development of a Bi2223 Insert Coil for a Conduction-Cooled 19 T Superconducting Magnet." *IEEE Transactions on Appiled Superconductivity* **15**, 1512–1515 (2005). http://dx.doi.org/10.1109/TASC.2005.849150. (Cited on page 2)

[TOO⁺11] Tomioka, A., Otonari, T., Ogata, T., Iwakuma, M., Okamoto, H., Hayashi, H., Iijima, Y., Saito, T., Gosho, Y., Tanabe, K., Izumi, T. and Shiohara, Y. "The short-circuit test results of 6.9kV/2.3kV 400kVA-class YBCO model transformer." *Physica C: Superconductivity* **471**, 1374–1378 (2011). http://dx.doi.org/10.1016/j.physc.2011.05.197. (Cited on page 1)

[TSO12] Terao, Y., Sekino, M. and Ohsaki, H. "Electromagnetic Design of 10 MW Class Fully Superconducting Wind Turbine Generators." *IEEE Transactions on Applied Superconductivity* **22**, 5201 904–5201 904 (2012). http://dx.doi.org/10.1109/TASC.2011.2177628. (Cited on page 1)

[Tsu08] Tsukamoto, O. "Overview of superconductivity in Japan – Strategy road map and R&D status." *Physica C: Superconductivity* **468**, 1101–1111 (2008). http://dx.doi.org/10.1016/j.physc.2008.05.199. (Cited on pages 2 and 134)

[TWP89] Talvacchio, J., Wagner, G. and Pohl, H. "$YBa_2Cu_3O_7$ films grown on epitaxial MgO buffer layers on sapphire." *Physica C: Superconductivity* **162-164**, 659–660 (1989). http://dx.doi.org/10.1016/0921-4534(89)91196-9. (Cited on page 2)

[TYM⁺07] Takigawa, H., Yumura, H., Masuda, T., Watanabe, M., Ashibe, Y., Itoh, H., Suzawa, C., Hirose, M., Yatsuka, K., Sato, K. and Isojima, S. "The installation and test results for Albany HTS cable project." *Physica C: Superconductivity* **463-465**, 1127–1131 (2007). http://dx.doi.org/10.1016/j.physc.2007.04.253. (Cited on pages 2 and 134)

[UAY⁺10]   Umemoto, K., Aizawa, K., Yokoyama, M., Yoshikawa, K., Kimura, Y., Izumi, M., Ohashi, K., Numano, M., Okumura, K., Yamaguchi, M., Gocho, Y. and Kosuge, E. "Development of 1 MW-class HTS motor for podded ship propulsion system." *Journal of Physics: Conference Series* **234**, 032060 (2010). http://dx.doi.org/10.1088/1742-6596/234/3/032060. (Cited on page 1)

[vdL09]   van der Laan, D.C. "$YBa_2Cu_3O_{7-\delta}$ coated conductor cabling for low ac-loss and high-field magnet applications." *Superconductor Science and Technology* **22**, 065013 (2009). http://dx.doi.org/10.1088/0953-2048/22/6/065013. (Cited on page 135)

[vdLLG11]   van der Laan, D.C., Lu, X.F. and Goodrich, L.F. "Compact $GdBa_2Cu_3O_{7-\delta}$ coated conductor cables for electric power transmission and magnet applications." *Superconductor Science and Technology* **24**, 042001 (2011). http://dx.doi.org/10.1088/0953-2048/24/4/042001. (Cited on page 135)

[vdLNM⁺13]   van der Laan, D.C., Noyes, P.D., Miller, G.E., Weijers, H.W. and Willering, G.P. "Characterization of a high-temperature superconducting conductor on round core cables in magnetic fields up to 20 T." *Superconductor Science and Technology* **26**, 045005 (2013). http://dx.doi.org/10.1088/0953-2048/26/4/045005. (Cited on page 135)

[vdLvEtH⁺01]   van der Laan, D.C., van Eck, H.J.N., ten Haken, B., Schwartz, J. and ten Kate, H.H.J. "Temperature and magnetic field dependence of the critical current of $Bi_2Sr_2Ca_2Cu_3O_x$ tape conductors." *IEEE Transactions on Applied Superconductivity* **11**, 3345–3348 (2001). http://dx.doi.org/10.1109/77.919779. (Cited on page 9)

[vG02]   Šouc, J. and Gömöry, F. "New approach to the ac loss measurement in the superconducting secondary circuit of an iron-core transformer." *Superconductor Science and Technology* **15**, 927–932 (2002). http://dx.doi.org/0953-2048/02/060927. (Cited on page 72)

[vGV05]   Šouc, J., Gömöry, F. and Vojenčiak, M. "Calibration free method for measurement of the AC magnetization loss." *Superconductor Science and Technology* **18**, 592–595 (2005). http://dx.doi.org/10.1088/0953-2048/18/5/003. (Cited on page 70)

[VKS07]   Vinod, K., Kumar, R.G.A. and Syamaprasad, U. "Prospects for MgB_2 superconductors for magnet application." *Superconductor Science and Technology* **20**, R1–R13 (2007). http://dx.doi.org/10.1088/0953-2048/20/1/R01. (Cited on page 23)

[VLW04]   Vagner, I.D., Lembrikov, B.I. and Wyder, P. *Electrodynamics of Magnetoactive Materials*. Springer-Verlag, Berlin, 1st ed. (2004). ISBN 3-540-43694-4. (Cited on page 18)

[Voj10]   Vojenčiak, M. "Influence of ferromagnetic material on performance of high temperature superconductor in form of tape." Ph.D. thesis, Slovak Academy of Sciences (2010). (Cited on page 69)

[WC52]   Wexler, A. and Corak, W. "Superconductivity of Vanadium." *Physical Review* **85**, 85–90 (1952). http://dx.doi.org/10.1103/PhysRev.85.85. (Cited on page 5)

[WFDR⁺12]   Werfel, F.N., Floegel-Delor, U., Rothfeld, R., Riedel, T., Goebel, B., Wippich, D. and Schirrmeister, P. "Superconductor bearings, flywheels and transportation." *Superconductor Science and Technology* **25**, 014 007 (2012). http://dx.doi.org/10.1088/0953-2048/25/1/014007. (Cited on page 2)

[WMV⁺13]   Weijers, H.W., Markiewicz, W.D., Voran, A.J., Gundlach, S.R., Sheppard, W.R., Jarvis, B., Noyes, P.D., Lu, J., Kandel, H., Bai, H., Gavrilin, A.V., Viouchkov, Y.L., Larbalestier, D.C. and Abraimov, D.V. "Progress in the Development of a Superconducting 32 T Magnet with REBCO High Field Coils." *IEEE Transactions on Applied Superconductivity* pp. 1–5 (2013). http://dx.doi.org/10.1109/TASC.2013.2288919. (Cited on page 2)

[WWS⁺09]   Wu, S., Wu, Y., Song, Y., Wu, W., Bi, Y., Xi, W., Xiao, L., Wang, Q., Ma, Y., Liu, X., Zhang, P., Xin, Y., Hou, B., Liu, R., Zhang, H., Han, Z., Zheng, J., Wang, J., Wang, S., Shi, J., Tang, Y., Qiu, M., Wei, B. and Tan, Y. "Recent Main Events in Applied Superconductivity in China." *IEEE Transactions on Applied Superconductivity* **19**, 1069–1080 (2009). http://dx.doi.org/10.1109/TASC.2009.2019244. (Cited on page 1)

[WZH⁺08]   Wang, Y., Zhao, X., Han, J., Li, H., Guan, Y., Bao, Q., Xu, X., Dai, S., Song, N., Zhang, F., Lin, L. and Xiao, L. "Development and test in grid of 630 kVA three-phase high temperature superconducting transformer." *Frontiers of Electrical and Electronic Engineering in China* **4**, 104–113 (2008). http://dx.doi.org/10.1007/s11460-009-0010-5. (Cited on page 1)

[XHB+04]   Xin, Y., Hou, B., Bi, Y., Cao, K., Zhang, Y., Wu, S., Ding, H., Wang, G., Liu, Q. and Han, Z. "China's 30 m, 35 kV/2 kA ac HTS power cable project." *Superconductor Science and Technology* **17**, S332–S335 (2004). http://dx.doi.org/10.1088/0953-2048/17/5/048. (Cited on pages 2 and 134)

[XLC+12]   Xu, Z., Lewin, R., Campbell, A.M., Cardwell, D.a. and Jones, H. "Simulation studies on the magnetization of (RE)BCO bulk superconductors using various split-coil arrangements." *Superconductor Science and Technology* **25**, 025 016 (2012). http://dx.doi.org/10.1088/0953-2048/25/2/025016. (Cited on page 2)

[XWD+08]   Xiao, L., Wang, Z., Dai, S., Zhang, J., Zhang, D., Gao, Z., Song, N. and Zhang, F. "Fabrication and Tests of a 1 MJ HTS Magnet for SMES." *IEEE Transactions on Applied Superconductivity* **18**, 770–773 (2008). http://dx.doi.org/10.1109/TASC.2008.922234. (Cited on page 1)

[YAI+09]   Yumura, H., Ashibe, Y., Itoh, H., Ohya, M., Watanabe, M., Masuda, T. and Weber, C. "Phase II of the Albany HTS Cable Project." *IEEE Transactions on Applied Superconductivity* **19**, 1698–1701 (2009). http://dx.doi.org/10.1109/TASC.2009.2017865. (Cited on pages 2 and 134)

[YHB+96]   Yang, Y., Hughes, T., Beduz, C., Spiller, D.M., Scurlock, R.G. and Norris, W.T. "The influence of geometry on self-field AC losses of Ag sheathed PbBi2223 tapes." *Physica C: Superconductivity* **256**, 378–386 (1996). http://dx.doi.org/10.1016/0921-4534(95)00662-1. (Cited on page 73)

[YHP+07]   Yeom, H., Hong, Y., Park, S., Seo, T., Seong, K. and Kim, H. "Study of Cryogenic Conduction Cooling Systems for an HTS SMES." *IEEE Transactions on Applied Superconductivity* **17**, 1955–1958 (2007). http://dx.doi.org/10.1109/TASC.2007.898491. (Cited on page 2)

[YKT+14]   Yamamoto, S., Konii, K., Tanabe, H., Yokoyama, S., Matsuda, T. and Yamada, T. "Super-Stable Superconducting MRI Magnet Operating for 25 Years." *IEEE Transactions on Applied Superconductivity* **24**, 1–4 (2014). http://dx.doi.org/10.1109/TASC.2013.2283228. (Cited on page 3)

[Zn13]   Zermeño, V.M.R. "On the Application of Fairy Dust: Sprinkling Meas-
              urement Data." *Proceedings of the Hogwarts School of Witchcraft and
              Wizardry: Advances in Divination* **1**, 1 (2013). (Cited on page 144)

[ZnGS13]   Zermeño, V.M.R., Grilli, F. and Sirois, F. "A full 3D time-dependent elec-
              tromagnetic model for Roebel cables." *Superconductor Science and Tech-
              nology* **26**, 052 001 (2013). http://dx.doi.org/10.1088/0953-2048/
              26/5/052001. (Cited on page 148)

A Appendix

A.1 FEM Sourcecode

Following is the Matlab source code in order to reproduce the FEM simulations using COMSOL.

A.1.1 Multipurpose H-model MATLAB/COMSOL Script

The following MATLAB script remotely controls COMSOL. Using this script, it is possible to simulate single coated conductors, coated conductor stacks, bifilar and pancake coils with and without ferromagnetic shielding. All parameters can be conveniently set in the beginning of the source code.

```
% function out = GeoSel_DoubleSweep

import com.comsol.model.*
import com.comsol.model.util.*

model = ModelUtil.create('Model');
ModelUtil.showProgress(true);

model.resetAuthor('Philipp_A._C._Krueger');

% Choose Geometry
geo = 1; % strip
% geo = 2; % substrate
% geo = 3; % shields
% geo = 4; % fcl
% geo = 5; % fcl w shields
% geo = 6; % pancake coil
% geo = 7; % pancake coil w shields
% geo = 8; % stack of 5 tapes
% geo = 9; % stack of 5 tapes, shielded
% geo = 10; % stack of 5 tapes, spaced by substrate
% geo = 11; % stack of 5 tapes, spaced by substrate, shielded

% Define parameters
% Fraction of critical field to apply
% Happl = [1e-3 2e-3 5e-3 1e-2 2e-2 5e-2 1e-1 2e-1 4e-1 6e-1 8e-1 1 1.5 2 2.5 3 3.5 4 4.5 5 5.5 6];
Happl = 0;
% Fraction of critical current to apply
Iappl = [0.01 0.03 0.05 0.07 0.1 0.2 0.3 0.4 0.5 0.6 0.7 0.8 0.9 1];
% Iappl = 0;
```

```matlab
% Geometry parameters
% c = [2.0e-3 3.0e-3];
c = 3.0e-3;
gap = 1e-4;
gapTop = 5e-6;
gapSide = 5e-6;
gapBottom = 5e-6;
gapTapes = 2e-5;
Ic = 300;
MeshCountScWidth=200;
tSC = 1e-6;
tSub = 1e-4;
tShieldSide = 3e-5;
tShieldFace = 3e-5;
wSC = 1e-2;
% Solution details
nu=50;
TimeStart = 0;
TimeStop = 1/nu;
TimeStep = TimeStop/100;
TimeList = ['range(',num2str(TimeStart),',',num2str(TimeStep),',',num2str(TimeStop),')'];

Jc = (Ic/(wSC*tSC));
Hc = Jc*tSC/pi;
mu0 = pi*4e-7;
Bc = mu0*Hc;

model.param.set('Bc', 'mu0*Hc');
%model.param.set('Bext', 'Happl*Bc*sin(2*pi*nu*t)/mu0');
model.param.set('Ec', '1e-4[V/m]');
model.param.set('Happl', Happl(1));
model.param.set('Hc', 'Jc*tSC/pi');
model.param.set('Iappl', Iappl(1));
model.param.set('Ic', [num2str(Ic),'[A]']);
%model.param.set('Iext', 'Iappl*Ic*sin(2*pi*nu*t)');
model.param.set('Jc', 'Ic/(wSC*tSC)');
model.param.set('mu0', [num2str(mu0),'[N/A^2]']);
model.param.set('n', '35');
model.param.set('nu', [num2str(nu),'[Hz]']);
model.param.set('rhoAir', '1[m/S]');
model.param.set('rhoFM', '1[m/S]');
model.param.set('tSC', [num2str(tSC),'[m]']);
model.param.set('tRampDuration', [num2str(1/nu),'[s]']);
model.param.set('wSC', [num2str(wSC),'[m]']);

model.modelNode.create('mod1');

model.func.create('rm1', 'Ramp');
model.func.create('an1', 'Analytic');
model.func.create('an2', 'Analytic');
model.func('rm1').set('smoothloc', true);
model.func('rm1').set('cutoffactive', true);
model.func('rm1').set('location', 'tRampDuration*2e-3');
```

```matlab
model.func('rm1').set('smoothzone', 'tRampDuration*3e-3');
model.func('rm1').set('slope', '40/tRampDuration');
model.func('an1').set('fununit', 'A');
model.func('an1').set('expr', 'rm1(t)*Iappl*Ic*sin(2*pi*nu*t)');
model.func('an1').set('argunit', 's');
model.func('an1').set('args', {'t'});
model.func('an1').set('plotargs', {'t' '0' 'tRampDuration'});
model.func('an1').set('funcname', 'Iext');
model.func('an2').set('fununit', 'A/m');
model.func('an2').set('expr', 'rm1(t)*Happl*Hc*sin(2*pi*nu*t)');
model.func('an2').set('argunit', 's');
model.func('an2').set('args', {'t'});
model.func('an2').set('plotargs', {'t' '0' 'tRampDuration'});
model.func('an2').set('funcname', 'Hext');

model.geom.create('geom1', 2);
model.geom('geom1').feature.create('c1', 'Circle');
model.geom('geom1').feature('c1').set('r', '0.5');
model.geom('geom1').feature.create('r1', 'Rectangle');
% r1: superconductor
model.geom('geom1').feature('r1').set('pos', {'0' '0'});
model.geom('geom1').feature('r1').set('base', 'center');
model.geom('geom1').feature('r1').set('size', {'wSC_[1/m]' 'tSC_[1/m]'});
if ((geo==3)||(geo==5)||(geo==7)||(geo==9)||(geo==11))
    if ((geo==3)||(geo==5)||(geo==7))
        ShieldWidth = num2str(c(1)+gapSide+tShieldSide,'%10.6e');
        ShieldThickness = num2str(2*tShieldFace+gapTop+gap+gapBottom+tSC+tSub,'%10.6e');
        CutWidth = num2str(wSC+2*gapSide,'%10.6e');
        CutThickness = num2str(tSC+gapTop+gap+gapBottom+tSub,'%10.6e');
        leftShieldPosX = num2str(-wSC/2-gapSide-tShieldSide+(c(1)+gapSide+tShieldSide)/2,'%10.6
            e');
        leftShieldPosY = num2str(-(gap+gapBottom+tShieldFace+tSC/2+tSub)+(2*tShieldFace+gapTop+
            gap+gapBottom+tSC+tSub)/2,'%10.6e');
        rightShieldPosX = num2str(wSC/2+gapSide+tShieldSide-(c(1)+gapSide+tShieldSide)/2,'%10.6
            e');
        rightShieldPosY = num2str(-(gap+gapBottom+tShieldFace+tSC/2+tSub)+(2*tShieldFace+gapTop
            +gap+gapBottom+tSC+tSub)/2,'%10.6e');
        carveBoxPosX = '0';
        carveBoxPosY = num2str(-(gap+gapBottom+tShieldFace+tSC/2+tSub)+(2*tShieldFace+gapTop+
            gap+gapBottom+tSC+tSub)/2,'%10.6e');
    elseif ((geo==8)||(geo==9))
        ShieldWidth = num2str(c(1)+gapSide+tShieldSide,'%10.6e');
        ShieldThickness = num2str(2*tShieldFace+gapTop+gapBottom+5*tSC+4*gapTapes,'%10.6e');
        CutWidth = num2str(wSC+2*gapSide,'%10.6e');
        CutThickness = num2str(5*tSC+4*gapTapes+gapTop+gapBottom,'%10.6e');
        leftShieldPosX = num2str(-wSC/2-gapSide-tShieldSide+(c(1)+gapSide+tShieldSide)/2,'%10.6
            e');
        leftShieldPosY = '0';
        rightShieldPosX = num2str(wSC/2+gapSide+tShieldSide-(c(1)+gapSide+tShieldSide)/2,'%10.6
            e');
        rightShieldPosY = '0';
        carveBoxPosX = '0';
        carveBoxPosY = '0';
```

```matlab
    elseif ((geo==11))
            ShieldWidth = num2str(c(1)+gapSide+tShieldSide,'%10.6e');
            ShieldThickness = num2str(5*(tSC+gap+tSub)+4*gapTapes+gapTop+gapBottom+2*tShieldFace,'
                %10.6e');
            CutWidth = num2str(wSC+2*gapSide,'%10.6e');
            CutThickness = num2str(5*(tSC+gap+tSub)+4*gapTapes+gapTop+gapBottom,'%10.6e');
            leftShieldPosX = num2str((c(1)+gapSide+tShieldSide-wSC)/2-gapSide-tShieldSide,'%10.6e')
                ;
            leftShieldPosY = num2str(tSC/2+gapTop+tShieldFace-(5*(tSC+gap+tSub)+4*gapTapes+gapTop+
                gapBottom+2*tShieldFace)/2,'%10.6e');
            rightShieldPosX = num2str(gapSide+tShieldSide+(wSC-(c(1)+gapSide+tShieldSide))/2,'%10.6
                e');
            rightShieldPosY = num2str(tSC/2+tShieldFace+gapTop-(5*(tSC+gap+tSub)+4*gapTapes+
                gapBottom+2*tShieldFace)/2,'%10.6e');
            carveBoxPosX = '0';
            carveBoxPosY = num2str(tSC/2+gapTop+tShieldFace-(5*(tSC+gap+tSub)+4*gapTapes+gapTop+
                gapBottom+2*tShieldFace)/2,'%10.6e');
    end
    model.geom('geom1').feature.create('r2', 'Rectangle');
    model.geom('geom1').feature.create('r3', 'Rectangle');
    model.geom('geom1').feature.create('r4', 'Rectangle');
    model.geom('geom1').feature.create('dif1', 'Difference');
    % r2: left shield
    model.geom('geom1').feature('r2').set('pos', {leftShieldPosX leftShieldPosY});
    model.geom('geom1').feature('r2').set('base', 'center');
    model.geom('geom1').feature('r2').set('size', {ShieldWidth ShieldThickness});
    model.geom('geom1').feature('r2').active(true);
    % r3: right shield
    model.geom('geom1').feature('r3').set('pos', {rightShieldPosX rightShieldPosY});
    model.geom('geom1').feature('r3').set('base', 'center');
    model.geom('geom1').feature('r3').set('size', {ShieldWidth ShieldThickness});
    model.geom('geom1').feature('r3').active(true);
    % r4: shield carve block
    model.geom('geom1').feature('r4').set('pos', {carveBoxPosX carveBoxPosY});
    model.geom('geom1').feature('r4').set('base', 'center');
    model.geom('geom1').feature('r4').set('size', {CutWidth CutThickness});
    model.geom('geom1').feature('r4').active(true);
    % dif1: carve shields
    model.geom('geom1').feature('dif1').selection('input').set({'r2' 'r3'});
    model.geom('geom1').feature('dif1').selection('input2').set({'r4'});
    model.geom('geom1').feature('dif1').active(true);
end
% carve FCL geometry slab
if ((geo==4)||(geo==5)||(geo==6)||(geo==7))
    model.geom('geom1').feature.create('r6', 'Rectangle');
    model.geom('geom1').feature.create('int1', 'Intersection');
    model.geom('geom1').feature('r6').active(true);
    model.geom('geom1').feature('r6').set('base', 'center');
    model.geom('geom1').feature('r6').set('size', {'0.3' '1.5e-3'});
    model.geom('geom1').feature('r6').set('pos', {'0' '0'});
    model.geom('geom1').feature('int1').selection('input').set({'c1' 'r6'});
    model.geom('geom1').feature('int1').active(true);
end
```

```matlab
if ((geo==8)||(geo==9))
      model.geom('geom1').feature.create('r7', 'Rectangle');
      model.geom('geom1').feature.create('r8', 'Rectangle');
      model.geom('geom1').feature.create('r9', 'Rectangle');
      model.geom('geom1').feature.create('r10', 'Rectangle');
      model.geom('geom1').feature('r7').set('pos', {'0' num2str(0-gapTapes-tSC/2,'%10.6e')});
      model.geom('geom1').feature('r7').set('base', 'center');
      model.geom('geom1').feature('r7').set('size', {'wSC' 'tSC'});
      model.geom('geom1').feature('r7').active(true);
      model.geom('geom1').feature('r8').set('pos', {'0' num2str(0-2*gapTapes-tSC/2,'%10.6e')});
      model.geom('geom1').feature('r8').set('base', 'center');
      model.geom('geom1').feature('r8').set('size', {'wSC' 'tSC'});
      model.geom('geom1').feature('r8').active(true);
      model.geom('geom1').feature('r9').set('pos', {'0' num2str(0+gapTapes-tSC/2,'%10.6e')});
      model.geom('geom1').feature('r9').set('base', 'center');
      model.geom('geom1').feature('r9').set('size', {'wSC' 'tSC'});
      model.geom('geom1').feature('r9').active(true);
      model.geom('geom1').feature('r10').set('pos', {'0' num2str(0+2*gapTapes-tSC/2,'%10.6e')});
      model.geom('geom1').feature('r10').set('base', 'center');
      model.geom('geom1').feature('r10').set('size', {'wSC' 'tSC'});
      model.geom('geom1').feature('r10').active(true);
elseif ((geo==10)||(geo==11))
      model.geom('geom1').feature.create('r7', 'Rectangle');
      model.geom('geom1').feature.create('r8', 'Rectangle');
      model.geom('geom1').feature.create('r9', 'Rectangle');
      model.geom('geom1').feature.create('r10', 'Rectangle');
      model.geom('geom1').feature('r7').set('pos', {'0' num2str(-tSC-gap-tSub-gapTapes,'%10.6e')});
      model.geom('geom1').feature('r7').set('base', 'center');
      model.geom('geom1').feature('r7').set('size', {'wSC' 'tSC'});
      model.geom('geom1').feature('r7').active(true);
      model.geom('geom1').feature('r8').set('pos', {'0' num2str(2*(-tSC-gap-tSub-gapTapes),'%10.6e')
          });
      model.geom('geom1').feature('r8').set('base', 'center');
      model.geom('geom1').feature('r8').set('size', {'wSC' 'tSC'});
      model.geom('geom1').feature('r8').active(true);
      model.geom('geom1').feature('r9').set('pos', {'0' num2str(3*(-tSC-gap-tSub-gapTapes),'%10.6e')
          });
      model.geom('geom1').feature('r9').set('base', 'center');
      model.geom('geom1').feature('r9').set('size', {'wSC' 'tSC'});
      model.geom('geom1').feature('r9').active(true);
      model.geom('geom1').feature('r10').set('pos', {'0' num2str(4*(-tSC-gap-tSub-gapTapes),'%10.6e'
          )});
      model.geom('geom1').feature('r10').set('base', 'center');
      model.geom('geom1').feature('r10').set('size', {'wSC' 'tSC'});
      model.geom('geom1').feature('r10').active(true);
elseif (geo==2)
   model.geom('geom1').feature.create('r2', 'Rectangle');
   %model.geom('geom1').feature('r2').set('pos', {'0' num2str(tSub/2+gap)});
   model.geom('geom1').feature('r2').set('pos', {'0' num2str(-tSC/2-gap-tSub/2)});
      model.geom('geom1').feature('r2').set('base', 'center');
      model.geom('geom1').feature('r2').set('size', {num2str(wSC) num2str(tSub)});
      model.geom('geom1').feature('r2').active(true);
end
```

```matlab
% force min. repair tolerance below smallest detail, else geometry will build uncontrollably
model.geom('geom1').feature('fin').set('repairtol', '1.0E-8');
model.geom('geom1').run;


% define superconducting domain
model.variable.create('var1');
model.variable('var1').model('mod1');
model.variable('var1').set('rho', 'Ec/Jc*abs(Jz/Jc)^(n-1)');
model.variable('var1').set('mu', 'mu0');
model.variable('var1').selection.geom('geom1', 2);
if ((geo==1)||(geo==4)||(geo==6))
      model.variable('var1').selection.set(2);
elseif ((geo==3)||(geo==5)||(geo==7))
      model.variable('var1').selection.set(3);
elseif ((geo==8)||(geo==10))
      model.variable('var1').selection.set([2 3 4 5 6]);
elseif ((geo==9)||(geo==11))
      model.variable('var1').selection.set([3 4 5 6 7]);
end


% define vacuum domain
model.variable.create('var2');
model.variable('var2').model('mod1');
model.variable('var2').set('rho', 'rhoAir');
model.variable('var2').set('mu', 'mu0');
model.variable('var2').selection.geom('geom1', 2);
model.variable('var2').selection.set(1);


% define general physics
model.variable.create('var3');
model.variable('var3').model('mod1');
model.variable('var3').set('Jz', 'H2x-H1y');
model.variable('var3').set('Ez', 'rho*Jz');
model.variable('var3').set('Hm', '((H1+0.001)^2+(H2+0.001)^2)^0.5');
model.variable('var3').set('Hmax', 'Iext(t)/(2*pi*(x^2+y^2)^0.5)');
model.variable('var3').set('Bm', 'mu*Hm');
% define ferromagnetic domain
if ((geo==3)||(geo==5)||(geo==7)||(geo==9)||(geo==11))
      model.variable.create('var4');
      model.variable('var4').model('mod1');
      model.variable('var4').set('rho', 'rhoFM');
      %%%%%%%%%%%%%%%%%%%%%%%%%%%%%%%%%%%%%%%%%%%%%%%%%%%%%%%%%%%%%%%%%%%%%%%%%%%%
      % Nguyen formula for local permeability mu=f(H(x,y)) (RABiTS)
      % model.variable('var4').set('muFM', '1+30600*(1-exp(-((Hm[m/A])/295)^2.5))/(Hm[m/A])^0.81+45*
          exp(-((Hm[m/A])/120)^2.5)');
      % model.variable('var4').set('dmuFM', '-24786*(1-exp(-((Hm[m/A])/295)^2.5))/(Hm[m/A])
          ^1.81+259.322*exp(-((Hm[m/A])/295)^2.5)*((Hm[m/A])/295)^1.5/(Hm[m/A])^0.81-0.9375*exp(-((
          Hm[m/A])/120)^2.5)*((Hm[m/A])/120)^1.5');
      % model.variable('var4').set('muH1X', '(muFM+dmuFM*H1^2/(Hm[m/A]))*mu0');
      % model.variable('var4').set('muH2Y', '(muFM+dmuFM*H2^2/(Hm[m/A]))*mu0');
      % model.variable('var4').set('muH12', 'mu0*H1[A/m]*H2[A/m]*dmuFM/Hm');
      % model.variable('var4').set('mu', '(muFM+dmuFM*Hm^2[m^2/A^2]/(Hm[m/A]))*mu0');
      %%%%%%%%%%%%%%%%%%%%%%%%%%%%%%%%%%%%%%%%%%%%%%%%%%%%%%%%%%%%%%%%%%%%%%%%%%%%%%%%%
```

```matlab
      % Constant mu (independent of H)
      % model.variable('var4').set('muH1', 'mu0');
      % model.variable('var4').set('muH2', 'mu0');
      %%%%%%%%%%%%%%%%%%%%%%%%%%%%%%%%%%%%%%%%%%%%%%%%%%%%%%%%%%%%%%%%%%%%%%
      % formula for local permeability based on Goemoery data mu=f(B(x,y)) (Ni) -> converted to mu(H
          (x,y))
      % model.variable('var4').set('muFM', '48.770*exp(-1.849e-4*(Hm[m/A])^8.267e-1)+69.969*exp
          (-4.214e-5*(Hm[m/A])^1.186)+1');
      % model.variable('var4').set('dmuFM', '-0.00745483*exp(-0.0001849*(Hm[m/A])^0.8267)/(Hm[m/A])
          ^0.1733-0.00349691*exp(-0.00004214*(Hm[m/A])^1.186)*(Hm[m/A])^0.186');
      % model.variable('var4').set('dmuFM', '0.012738/((1+2.89e-8*x**2)*atan(0.00017*x)**0.00756)');
      % model.variable('var4').set('muH1X', '(muFM+dmuFM*(H1[m/A])^2/(Hm[m/A]))*mu0');
      % model.variable('var4').set('muH2Y', '(muFM+dmuFM*(H2[m/A])^2/(Hm[m/A]))*mu0');
      % model.variable('var4').set('muH12', 'mu0*H1[m/A]*H2[m/A]*dmuFM/(Hm[m/A])');
      model.variable('var4').set('muFM', '-75.5*atan(0.00017*Hm[m/A])^0.99244+119.190934');
      model.variable('var4').set('mu', 'muFM*mu0');
      %%%%%%%%%%%%%%%%%%%%%%%%%%%%%%%%%%%%%%%%%%%%%%%%%%%%%%%%%%%%%%%%%%%%%%
      % Ni-W
      % model.variable('var4').set('muFM', '1+30600*(1-exp(-((Hm[m/A])/295)^2.5))/(Hm[m/A])^0.81+45*
          exp(-((Hm[m/A])/120)^2.5)');
      % model.variable('var4').set('mu', 'muFM*mu0');
      %%%%%%%%%%%%%%%%%%%%%%%%%%%%%%%%%%%%%%%%%%%%%%%%%%%%%%%%%%%%%%%%%%%%%%
      model.variable('var4').selection.geom('geom1', 2);
      if ((geo==3)||(geo==5)||(geo==7))
            model.variable('var4').selection.set([2 4]);
      elseif ((geo==9)||(geo==11))
            model.variable('var4').selection.set([2 8]);
      end
end
if (geo==2)
      model.variable('var1').selection.set(3);
      model.variable.create('var4');
      model.variable('var4').model('mod1');
      model.variable('var4').set('rho', 'rhoFM');
      %%%%%%%%%%%%%%%%%%%%%%%%%%%%%%%%%%%%%%%%%%%%%%%%%%%%%%%%%%%%%%%%%%%%%%
      % Nguyen formula for local permeability mu=f(H(x,y)) (RABiTS)
      % model.variable('var4').set('muFM', '-75.5*atan(0.00017*Hm[m/A])^0.99244+119.190934');
      % model.variable('var4').set('mu', 'mu0*muFM');
      %%%%%%%%%%%%%%%%%%%%%%%%%%%%%%%%%%%%%%%%%%%%%%%%%%%%%%%%%%%%%%%%%%%%%%
      % Constant mu (independent of H)
      % model.variable('var4').set('muH1', 'mu0');
      % model.variable('var4').set('muH2', 'mu0');
      %%%%%%%%%%%%%%%%%%%%%%%%%%%%%%%%%%%%%%%%%%%%%%%%%%%%%%%%%%%%%%%%%%%%%%
      % Ni-W
      model.variable('var4').set('muFM', '1+30600*(1-exp(-((Hm[m/A])/295)^2.5))/(Hm[m/A])^0.81+45*
          exp(-((Hm[m/A])/120)^2.5)');
      model.variable('var4').set('mu', 'muFM*mu0');
      %%%%%%%%%%%%%%%%%%%%%%%%%%%%%%%%%%%%%%%%%%%%%%%%%%%%%%%%%%%%%%%%%%%%%%
      model.variable('var4').selection.geom('geom1', 2);
      model.variable('var4').selection.set(2);
end

model.physics.create('g', 'GeneralFormPDE', 'geom1');
```

```matlab
model.physics('g').feature.create('dir1', 'DirichletBoundary', 1);
if (geo==1)
        model.physics('g').feature('dir1').selection.set([5 6 7 8]);
elseif (geo==2)
   model.physics('g').feature('dir1').selection.set([9 10 11 12]);
elseif (geo==3)
        model.physics('g').feature('dir1').selection.set([21 22 23 24]);
elseif ((geo==4)||(geo==5)||(geo==6)||(geo==7))
        model.physics('g').feature('dir1').selection.set([2 3]);
elseif ((geo==8)||(geo==10))
        model.physics('g').feature('dir1').selection.set([21 22 23 24]);
elseif ((geo==9)||(geo==11))
        model.physics('g').feature('dir1').selection.set([37 38 39 40]);
end
% for FCL geometry to force symetrical boundary conditions
% model.physics('g').feature('dir1').selection.set([2 3]);
model.physics('g').feature.create('constr1', 'PointwiseConstraint', 1);
model.physics('g').feature('constr1').selection.set(1);
if (geo==4)
        model.physics('g').feature.create('constr2', 'PointwiseConstraint', 1);
        model.physics('g').feature('constr2').selection.set([2 3]);
end

model.mesh.create('mesh1', 'geom1');
model.mesh('mesh1').feature.create('map1', 'Map');
model.mesh('mesh1').feature('map1').selection.geom('geom1', 2);
if ((geo==1)||(geo==4)||(geo==6))
        model.mesh('mesh1').feature('map1').selection.set(2);
elseif (geo==2)
   model.mesh('mesh1').feature('map1').selection.set([2 3]);
elseif ((geo==3)||(geo==5)||(geo==7))
        model.mesh('mesh1').feature('map1').selection.set(3);
elseif ((geo==8)||(geo==10))
        model.mesh('mesh1').feature('map1').selection.set([2 3 4 5 6]);
elseif ((geo==9)||(geo==11))
        model.mesh('mesh1').feature('map1').selection.set([3 4 5 6 7]);
end
model.mesh('mesh1').feature('map1').feature.create('dis1', 'Distribution');
if (geo==1)
        model.mesh('mesh1').feature('map1').feature('dis1').selection.set([2 3]);
elseif (geo==2)
        model.mesh('mesh1').feature('map1').feature('dis1').selection.set([2 3 5 6]);
elseif (geo==3)
        model.mesh('mesh1').feature('map1').feature('dis1').selection.set([8 9]);
elseif ((geo==4)||(geo==6))
   model.mesh('mesh1').feature('map1').feature('dis1').selection.set([5 6]);
elseif ((geo==5)||(geo==7))
   model.mesh('mesh1').feature('map1').feature('dis1').selection.set([11 12]);
elseif ((geo==8)||(geo==10))
        model.mesh('mesh1').feature('map1').feature('dis1').selection.set([2 3 5 6 8 9 11 12 14 15]);
elseif ((geo==9)||(geo==11))
        model.mesh('mesh1').feature('map1').feature('dis1').selection.set([8 9 11 12 14 15 17 18 20
            21]);
```

```matlab
end
model.mesh('mesh1').feature('map1').feature.create('dis2', 'Distribution');
if (geo==1)
        model.mesh('mesh1').feature('map1').feature('dis2').selection.set([1 4]);
elseif (geo==2)
        model.mesh('mesh1').feature('map1').feature('dis2').selection.set(4);
    model.mesh('mesh1').feature('map1').feature.create('dis3', 'Distribution');
    model.mesh('mesh1').feature('map1').feature('dis3').selection.set(1);
elseif (geo==3)
        model.mesh('mesh1').feature('map1').feature('dis2').selection.set([7 18]);
elseif ((geo==4)||(geo==6))
    model.mesh('mesh1').feature('map1').feature('dis2').selection.set([4 7]);
elseif ((geo==5)||(geo==7))
    model.mesh('mesh1').feature('map1').feature('dis2').selection.set([10 21]);
elseif ((geo==8)||(geo==10))
    model.mesh('mesh1').feature('map1').feature('dis2').selection.set([1 4 7 10 13 16 17 18 19 20]);
elseif ((geo==9)||(geo==11))
    model.mesh('mesh1').feature('map1').feature('dis2').selection.set([7 10 13 16 19 30 31 32 33 34]);

end
model.mesh('mesh1').feature.create('ftri1', 'FreeTri');
if ((geo==4)||(geo==5)||(geo==6)||(geo==7))
    model.mesh('mesh1').feature('ftri1').feature.create('dis1', 'Distribution');
    model.mesh('mesh1').feature('ftri1').feature('dis1').selection.set([2 3]);
end
if ((geo==5)||(geo==7))
    model.mesh('mesh1').feature('ftri1').feature.create('dis2', 'Distribution');
    model.mesh('mesh1').feature('ftri1').feature('dis2').selection.set(9);
    model.mesh('mesh1').feature('ftri1').feature.create('dis3', 'Distribution');
    model.mesh('mesh1').feature('ftri1').feature('dis3').selection.set(19);
end


model.cpl.create('intop1', 'Integration', 'geom1');
if ((geo==1)||(geo==4)||(geo==6))
        model.cpl('intop1').selection.set(2);
elseif ((geo==2)||(geo==3)||(geo==5)||(geo==7))
        model.cpl('intop1').selection.set(3);
elseif ((geo==8)||(geo==10))
        model.cpl('intop1').selection.set([2 3 4 5 6]);
elseif ((geo==9)||(geo==11))
        model.cpl('intop1').selection.set([3 4 5 6 7]);
end


model.variable('var1').name('Domain:_Supercond');
model.variable('var2').name('Domain:_Vacuum');
model.variable('var3').name('Entire_Sphere');
if ((geo==3)||(geo==5)||(geo==7)||(geo==9))
        model.variable('var4').name('Domain:_Ferromagnet');
end


model.physics('g').field('dimensionless').component({'H1' 'H2'});
model.physics('g').field('dimensionless').field('H');
model.physics('g').prop('ShapeProperty').set('shapeFunctionType', 'shcurl');
```

```matlab
model.physics('g').prop('ShapeProperty').set('order', '1');
model.physics('g').prop('Units').set('DependentVariableQuantity', 'magneticfield');
model.physics('g').prop('Units').set('CustomSourceTermUnit', 1, 'V*m^-2');
model.physics('g').feature('gfeq1').set('f', {'-d(mu,t)*H1'; '-d(mu,t)*H2'});
model.physics('g').feature('gfeq1').set('Ga', {'0' 'Ez'; '-Ez' '0'});
%model.physics('g').feature('gfeq1').set('da', {'muH1X'; 'muH12'; 'muH12'; 'muH2Y'});
model.physics('g').feature('gfeq1').set('da', {'mu'; '0'; '0'; 'mu'});
if ((geo==1)||(geo==2)||(geo==3)||(geo==8)||(geo==9)||(geo==10)||(geo==11))
      model.physics('g').feature('dir1').set('r', {'-Hmax*y/(x^2+y^2)^0.5'; 'Hext(t)+Hmax*x/(x^2+y
          ^2)^0.5'});
elseif ((geo==4)||(geo==5)) % for FCL geometry to force symetrical boundary conditions
      model.physics('g').feature('dir1').set('r', {'0'; '0'});
      model.physics('g').feature('dir1').set('useDirichletCondition', 1, '0');
elseif ((geo==6)||(geo==7)) % for pancake geometry to force symetrical boundary conditions
      model.physics('g').feature('dir1').set('r', {'0'; '0'});
      model.physics('g').feature('dir1').set('useDirichletCondition', 1, '1');
      model.physics('g').feature('dir1').set('useDirichletCondition', 2, '0');
end
model.physics('g').feature('dir1').active(true);
model.physics('g').feature('constr1').set('constraintExpression', 'intop1(Jz)-Iext(t)');
model.physics('g').feature('constr1').active(true);


model.mesh('mesh1').feature('size').set('hauto', 6);
model.mesh('mesh1').feature('map1').feature('dis1').set('type', 'predefined');
model.mesh('mesh1').feature('map1').feature('dis1').set('elemcount', MeshCountScWidth);
model.mesh('mesh1').feature('map1').feature('dis1').set('elemratio', '8');
model.mesh('mesh1').feature('map1').feature('dis1').set('method', 'geometric');
model.mesh('mesh1').feature('map1').feature('dis1').set('symmetric', true);
model.mesh('mesh1').feature('map1').feature('dis2').set('numelem', '1');
if (geo==2)
   model.mesh('mesh1').feature('map1').feature('dis3').set('numelem', '10');
end
if ((geo==4)||(geo==5)||(geo==6)||(geo==7))
      model.mesh('mesh1').feature('ftri1').feature('dis1').set('type', 'predefined');
      model.mesh('mesh1').feature('ftri1').feature('dis1').set('elemcount', MeshCountScWidth*2);
      model.mesh('mesh1').feature('ftri1').feature('dis1').set('elemratio', '16');
      model.mesh('mesh1').feature('ftri1').feature('dis1').set('method', 'geometric');
      model.mesh('mesh1').feature('ftri1').feature('dis1').set('symmetric', true);
      model.mesh('mesh1').feature('ftri1').feature('dis1').set('reverse', 'on');
end
if ((geo==5)||(geo==7))
      model.mesh('mesh1').feature('ftri1').feature('dis2').set('type', 'predefined');
      model.mesh('mesh1').feature('ftri1').feature('dis2').set('elemcount', MeshCountScWidth/(wSC/c)
          +2);
      model.mesh('mesh1').feature('ftri1').feature('dis2').set('elemratio', '8');
      model.mesh('mesh1').feature('ftri1').feature('dis2').set('method', 'geometric');
      model.mesh('mesh1').feature('ftri1').feature('dis2').set('symmetric', false);
      model.mesh('mesh1').feature('ftri1').feature('dis2').set('reverse', 'off');
      model.mesh('mesh1').feature('ftri1').feature('dis3').set('type', 'predefined');
      model.mesh('mesh1').feature('ftri1').feature('dis3').set('elemcount', MeshCountScWidth/(wSC/c)
          +2);
      model.mesh('mesh1').feature('ftri1').feature('dis3').set('elemratio', '8');
      model.mesh('mesh1').feature('ftri1').feature('dis3').set('method', 'geometric');
```

```matlab
        model.mesh('mesh1').feature('ftri1').feature('dis3').set('symmetric', false);
        model.mesh('mesh1').feature('ftri1').feature('dis3').set('reverse', 'on');
end
model.mesh('mesh1').run;

model.study.create('std1');
model.study('std1').feature.create('time', 'Transient');

model.sol.create('sol1');
model.sol('sol1').study('std1');
model.sol('sol1').attach('std1');
model.sol('sol1').feature.create('st1', 'StudyStep');
model.sol('sol1').feature.create('v1', 'Variables');
model.sol('sol1').feature.create('t1', 'Time');

model.result.dataset.create('cln1', 'CutLine2D');
model.result.create('pg1', 'PlotGroup2D');
model.result('pg1').feature.create('surf1', 'Surface');
model.result.create('pg2', 'PlotGroup1D');
model.result('pg2').feature.create('lngr1', 'LineGraph');
model.result.create('pg3', 'PlotGroup1D');
model.result('pg3').feature.create('lngr1', 'LineGraph');
model.result.create('pg4', 'PlotGroup1D');
model.result('pg4').feature.create('ptgr1', 'PointGraph');

model.study('std1').feature('time').set('tlist', TimeList);
model.study('std1').feature('time').set('rtol', '1e-7');
model.study('std1').feature('time').set('rtolactive', false);
model.study('std1').feature('time').set('plot', 'off');

model.sol('sol1').feature('st1').name('Compile_Equations:_Time_Dependent');
model.sol('sol1').feature('st1').set('studystep', 'time');
model.sol('sol1').feature('v1').set('control', 'time');
model.sol('sol1').feature('t1').set('control', 'time');
model.sol('sol1').feature('t1').set('tlist', 'range(0,1/nu/100,1/nu)');
model.sol('sol1').feature('t1').set('rtol', '1e-7');
model.sol('sol1').feature('t1').set('atolglobalmethod', 'unscaled');
model.sol('sol1').feature('t1').set('atolglobal', '1e-7');
model.sol('sol1').feature('t1').set('initialstepbdfactive', false);
model.sol('sol1').feature('t1').set('maxstepbdf', 'tRampDuration/1000');
model.sol('sol1').feature('t1').set('maxstepbdfactive', false);
model.sol('sol1').feature('t1').set('plot', 'off');
model.sol('sol1').feature('t1').feature('dDef').set('mumpsalloc', '4');
model.sol('sol1').feature('t1').feature('fcDef').active(true);
model.sol('sol1').feature('t1').feature('fcDef').set('niter', '25');
model.sol('sol1').feature('t1').feature('fcDef').set('ntermauto', 'itertol');
model.sol('sol1').feature('t1').feature('fcDef').set('dtech', 'auto');
model.sol('sol1').feature('t1').feature('aDef').set('nullfun', 'flspnull');
model.sol('sol1').feature('t1').feature('dDef').set('errorchk', 'off');


disp(['Geometry_sweep_(of_',num2str(length(c)),'):_1'])
disp('First_run!')
```

```matlab
if ~(exist(['Current_',num2str(Iappl(1)*Ic,'%1.5f'),'/','Field_',num2str(Happl(1)*Bc,'%1.5f'),'/','
    Happl_',num2str(Happl(1)*Bc,'%1.5f'),'_SCLossProfile.txt'],'file'))
        model.sol('sol1').runAll;
else
        model.param.set('Iappl', 0);
        model.param.set('Happl', 0);
        model.sol('sol1').runAll;
end

model.result.dataset('cln1').name('Supercond');
model.result.dataset('cln1').set('genpoints', {num2str(-wSC/2,'%10.6e') '0'; num2str(wSC/2,'%10.6e')
    '0'});
model.result('pg1').set('solnum', '76');
model.result('pg1').set('view', 'view1');
model.result('pg1').set('title', 'Local_magnetic_field_B_(@t=0.0076)');
model.result('pg1').set('windowtitle', 'Graphics');
model.result('pg1').set('titleactive', false);
%model.result('pg1').feature('surf1').set('rangecoloractive', 'on');
%model.result('pg1').feature('surf1').set('rangecolormin', '-10000');
%model.result('pg1').feature('surf1').set('rangecolormax', '10000');
model.result('pg2').name('1D_Plot_Jz');
model.result('pg2').set('data', 'cln1');
model.result('pg2').set('innerinput', 'manual');
model.result('pg2').set('solnum', {'76'});
model.result('pg2').set('title', 'local_current_density_Jz_(A/m^2)');
model.result('pg2').set('xlabel', 'x_coordinate_(m)');
model.result('pg2').set('ylabel', '-Jz_(A/m^2)');
model.result('pg2').feature('lngr1').set('expr', '-Jz/Jc');
model.result('pg2').feature('lngr1').set('unit', '1');
model.result('pg2').feature('lngr1').set('descr', '-Jz/Jc');
model.result('pg2').feature('lngr1').set('xdata', 'expr');
model.result('pg2').feature('lngr1').set('xdataexpr', 'x');
model.result('pg3').name('1D_Plot_B');
model.result('pg3').set('data', 'cln1');
model.result('pg3').set('innerinput', 'manual');
model.result('pg3').set('solnum', {'76'});
model.result('pg3').set('title', 'local_magnetic_field_(H/m)');
model.result('pg3').set('xlabel', 'x_coordinate_(m)');
model.result('pg2').set('ylabel', 'mu0*H2_(H/m)');
model.result('pg3').feature('lngr1').set('expr', 'mu0*H2');
model.result('pg3').feature('lngr1').set('unit', 'V/m');
model.result('pg3').feature('lngr1').set('descr', '');
model.result('pg3').feature('lngr1').set('xdata', 'expr');
model.result('pg3').feature('lngr1').set('xdataexpr', 'x');
model.result('pg4').set('title', 'Superconductor_ac-losses_P_(W/m)');
model.result('pg4').set('xlabel', 'Time_(s)');
model.result('pg4').set('ylabel', 'Losses_P_(W/m)');
model.result('pg4').feature('ptgr1').set('expr', 'intop1(Jz*Ez)');
model.result('pg4').feature('ptgr1').set('unit', 'W/m');
model.result('pg4').feature('ptgr1').set('descr', 'intop1(Jz*Ez)');
model.result('pg4').feature('ptgr1').selection.all;

model.result.export.create('img1', 'Image2D');
```

```matlab
model.result.export('img1').set('plotgroup', 'pg1');
model.result.export.create('img2', 'Image1D');
model.result.export('img2').set('plotgroup', 'pg2');
model.result.export.create('img3', 'Image1D');
model.result.export('img3').set('plotgroup', 'pg3');
model.result.export.create('img4', 'Image1D');
model.result.export('img4').set('plotgroup', 'pg4');


model.result.export.create('plot1', 'pg2', 'lngr1', 'Plot');
model.result.export.create('plot2', 'pg3', 'lngr1', 'Plot');
model.result.export.create('plot3', 'pg4', 'ptgr1', 'Plot');


%%% Start outer iteration
SCLossX=cell(length(Iappl),length(Happl)); % initialise matrix for speed
SCLoss=cell(length(Iappl),length(Happl)); % initialise matrix for speed
FMLoss=cell(length(Iappl),length(Happl)); % initialise matrix for speed
FMLossShieldLeft=cell(length(Iappl),length(Happl)); % initialise matrix for speed
FMLossShieldRight=cell(length(Iappl),length(Happl)); % initialise matrix for speed
for i=1:length(Iappl)
    for j=1:length(Happl)
        % skip if configuration has already been run
        CDir=['Current_',num2str(Iappl(i)*Ic,'%1.5f')]; % construct path from parameters
        FDir=['Field_',num2str(Happl(j)*Bc,'%1.5f')]; % construct path from parameters
        if (exist([CDir,'/',FDir,'/','Happl_',num2str(Happl(j)*Bc,'%1.5f'),'_SCLossProfile.txt'
            ],'file')) % skip iteration if directory already exists
                disp(['Iteration:_',num2str((i-1)*length(Happl)+j),'/',num2str(length(Happl)*
                    length(Iappl))])
                disp([CDir,'/',FDir,'/','Happl_',num2str(Happl(j)*Bc,'%1.5f'),'_SCLossProfile.
                    txt','_already_exists,_skipping...'])
                continue
        end
        % skip for first run because we want post-processing before starting next iteration
        if ~((i==1)&&(j==1))
                disp(['Iteration:_',num2str((i-1)*length(Happl)+j),'/',num2str(length(Happl)*
                    length(Iappl))])
                disp(['Now_running_simulation_with_parameters:_Iappl_',num2str(Iappl(i)*Ic),'_A_
                    and_Happl_',num2str(Happl(j)*Bc),'_T'])
                model.param.set('Iappl', Iappl(i)); % reset parameter for applied current Iappl
                model.param.set('Happl', Happl(j)); % reset parameter for applied field Happl
                model.sol('sol1').run; % run simulation
        end
        disp(['Simulation_solved_successfully_for_Iappl:_',num2str(Iappl(i)*Ic,'%1.5f'),',_A_
            and_Happl:_',num2str(Happl(j)*Bc,'%1.5f'),'_T'])


        % extract instantaneous SC losses for each timestep from SC domain in W/m
    tmpEval=mpheval(model,'intop1(Jz*Ez)','Edim',1,'Selection',1); % evaluate the integration
        intop1(Jz*Ez) (defined above) over all the SCs
        tmpEval.d1(:,1)=[]; % delete the superfluous first column that MATLAB adds to the
            matrix it being just a copy of the second
        % construct a matrix with the first column being the time and the
        % second the instantaneous losses:
    SCLossX{i,j}=[[TimeStart:TimeStep:TimeStop]' tmpEval.d1]; %#ok
```

```matlab
        % integrate SC losses over whole cycle and convert to J/m (divide by frequency nu)
        SCLoss{i,j}=1/nu*trapz(SCLossX{i,j}(ceil(length(SCLossX{i,j})/2):length(SCLossX{i,j})
            ,1),SCLossX{i,j}(ceil(length(SCLossX{i,j})/2):length(SCLossX{i,j}),2))/(SCLossX{i,
            j}(length(SCLossX{i,j}),1)-SCLossX{i,j}(ceil(length(SCLossX{i,j})/2),1));

        % extract FM losses in each shield
        if ((geo==3)||(geo==5)||(geo==7))
            FMLossShieldLeft{i}=mphint(model,'(2750[J/m^3]*(mu0*muFM*Hm/(0.5[T]))^2)*((mu0*
                muFM*Hm)<=0.4[T])+2750[J/m^3]*((mu0*muFM*Hm)>0.4[T])','t',SCLossX{i,j}(ceil(
                length(SCLossX{i,j})*3/4),1),'Edim',2,'selection',2);
            FMLossShieldRight{i}=mphint(model,'(2750[J/m^3]*(mu0*muFM*Hm/(0.5[T]))^2)*((mu0*
                muFM*Hm)<=0.4[T])+2750[J/m^3]*((mu0*muFM*Hm)>0.4[T])','t',SCLossX{i,j}(ceil(
                length(SCLossX{i,j})*3/4),1),'Edim',2,'selection',4);
elseif ((geo==9)||(geo==11))
            FMLossShieldLeft{i}=mphint(model,'(2750[J/m^3]*(mu0*muFM*Hm/(0.5[T]))^2)*((mu0*
                muFM*Hm)<=0.4[T])+2750[J/m^3]*((mu0*muFM*Hm)>0.4[T])','t',SCLossX{i,j}(ceil(
                length(SCLossX{i,j})*3/4),1),'Edim',2,'selection',2);
            FMLossShieldRight{i}=mphint(model,'(2750[J/m^3]*(mu0*muFM*Hm/(0.5[T]))^2)*((mu0*
                muFM*Hm)<=0.4[T])+2750[J/m^3]*((mu0*muFM*Hm)>0.4[T])','t',SCLossX{i,j}(ceil(
                length(SCLossX{i,j})*3/4),1),'Edim',2,'selection',8);
        end
        % combine for total FM loss
        if ((geo==1)||(geo==4)||(geo==6)||(geo==8))
            FMLoss{i,j}=0;
elseif (geo==2)
    FMLoss{i,j}=mphint(model,'(0.8[J/m^3]*(mu0*muFM*Hm/(0.01[T]))^1.87)','t',SCLossX{i,j}(ceil
        (length(SCLossX{i,j})*3/4),1),'Edim',2,'selection',2);
        elseif ((geo==3)||(geo==5)||(geo==7)||(geo==9)||(geo==11))
            FMLoss{i,j}=FMLossShieldLeft{i}+FMLossShieldRight{i};
        end

        % extract magnetic field strength close to boundary
        % Field{i}=abs(mphinterp(model,{'mu0*H2'},'coord',[0; 0.49],'t',SCLossX{i,j}(ceil(
            length(SCLossX{i,j})*3/4),1))); %#ok

        % save post-processed files
        if ~(exist(CDir,'dir'))
            mkdir(CDir);
        end
        cd(CDir);
            if ~(exist(FDir,'dir'))
                mkdir(FDir);
            end
            cd(FDir);
                fid=fopen(['Happl_',num2str(Happl(j)*Bc,'%1.5f'),'_SCLossProfile.txt'],'
                    wt');
                    fprintf(fid,'%2.5e_%2.5e\n',[SCLossX{i,j}(:,1) SCLossX{i,j}(:,2)
                        ;]');
                fclose(fid);
                fid=fopen(['Happl_',num2str(Happl(j)*Bc,'%1.5f'),'_Losses.txt'],'wt');
                    fprintf(fid,'%2.5e\t%2.5e\t%2.5e\t%2.5e\n',[Iappl(i)*Ic Happl(j)*
                        Bc SCLoss{i,j} FMLoss{i,j}]');
                fclose(fid);
```

```matlab
                model.result('pg1').run;
                model.result.export('img1').set('pngfilename', [pwd,'\','Happl_',num2str(
                    Happl(j)*Bc,'%1.5f'),'_2D_H1.png']);
                model.result.export('img1').set('options', 'on');
                model.result.export('img1').set('title', 'on');
                model.result.export('img1').set('logo', 'off');
                model.result.export('img1').set('legend', 'off');
                model.result.export('img1').run;
                model.result('pg2').run;
                model.result.export('img2').set('pngfilename', [pwd,'\','Happl_',num2str(
                    Happl(j)*Bc,'%1.5f'),'_1D_Jz.png']);
                model.result.export('img2').set('options', 'on');
                model.result.export('img2').set('title', 'on');
                model.result.export('img2').set('logo', 'off');
                model.result.export('img2').set('legend', 'off');
                model.result.export('img2').run;
                model.result('pg3').run;
                model.result.export('img3').set('pngfilename', [pwd,'\','Happl_',num2str(
                    Happl(j)*Bc,'%1.5f'),'_1D_B.png']);
                model.result.export('img3').set('options', 'on');
                model.result.export('img3').set('title', 'on');
                model.result.export('img3').set('logo', 'off');
                model.result.export('img3').set('legend', 'off');
                model.result.export('img3').run;
                model.result('pg4').run;
                model.result.export('img4').set('pngfilename', [pwd,'\','Happl_',num2str(
                    Happl(j)*Bc,'%1.5f'),'_1D_P.png']);
                model.result.export('img4').set('options', 'on');
                model.result.export('img4').set('title', 'on');
                model.result.export('img4').set('logo', 'off');
                model.result.export('img4').set('legend', 'off');
                model.result.export('img4').run;
                model.result.export('plot1').set('filename', [pwd,'\','Happl_',num2str(
                    Happl(j)*Bc,'%1.5f'),'_1D_Jz.txt']);
                model.result.export('plot1').run;
                model.result.export('plot2').set('filename', [pwd,'\','Happl_',num2str(
                    Happl(j)*Bc,'%1.5f'),'_1D_B.txt']);
                model.result.export('plot2').run;
                model.result.export('plot3').set('filename', [pwd,'\','Happl_',num2str(
                    Happl(j)*Bc,'%1.5f'),'_1D_P.txt']);
                model.result.export('plot3').run;
                model.save([pwd,'\','Happl_',num2str(Happl(j)*Bc,'%1.5f'),'_COMSOL.mph']);

        cd('..');
        % save Loss.txt for subsweep after each run in case a simulation crashes
        fid=fopen('Loss.txt','wt');
                fprintf(fid,'#Iappl\t\tHappl*Bc\t\tSCloss\t\t\tFMLoss\n');
                for n=1:j
                        fprintf(fid,'%2.5e\t%2.5e\t%2.5e\t%2.5e\n',[Iappl(i)*Ic Happl(n)*
                            Bc SCLoss{i,n} FMLoss{i,n}]');
                end
        fclose(fid);
    cd('..');
```

```
        end
end

% out = model;
```

A.1.2 High magnetic field coil with L-shaped ferromagnetic shields

```
import com.comsol.model.*
import com.comsol.model.util.*

model = ModelUtil.create('Model');

model.param.set('Bc', 'mu0*Hc');
model.param.set('Ec', '1e-4[V/m]');
model.param.set('Happl', '0');
model.param.set('Hc', 'Jc*tSC/pi');
model.param.set('Iappl', '1.1');
model.param.set('Ic', '300[A]');
model.param.set('Jc', 'Ic/(wSC*tSC)');
model.param.set('mu0', '1.2566e-06[N/A^2]');
model.param.set('n', '35');
model.param.set('nu', '50[Hz]');
model.param.set('rhoAir', '1[m/S]');
model.param.set('rhoFM', '1[m/S]');
model.param.set('tSC', '1e-6[m]');
model.param.set('tRampDuration', '10[s]');
model.param.set('wSC', '4e-3[m]');

model.modelNode.create('mod1');

model.func.create('rm1', 'Ramp');
model.func.create('an1', 'Analytic');
model.func.create('an2', 'Analytic');
model.func('rm1').set('slope', '1/tRampDuration');
model.func('rm1').set('smoothzone', 'tRampDuration*1e-2');
model.func('rm1').set('location', 'tRampDuration*1e-2');
model.func('rm1').set('cutoffactive', true);
model.func('rm1').set('smoothloc', true);
model.func('an1').set('fununit', 'A');
model.func('an1').set('plotargs', {'t' '0' 'tRampDuration'});
model.func('an1').set('funcname', 'Iext');
model.func('an1').set('args', {'t'});
model.func('an1').set('argunit', 's');
model.func('an1').set('expr', 'rm1(t)*Iappl*Ic');
model.func('an2').set('fununit', 'A/m');
model.func('an2').set('plotargs', {'t' '0' 'tRampDuration'});
model.func('an2').set('funcname', 'Hext');
model.func('an2').set('args', {'t'});
model.func('an2').set('argunit', 's');
model.func('an2').set('expr', 'rm1(t)*mu0*Happl*Hc*sin(2*pi*nu*t)');
```

```
model.geom.create('geom1', 2);
model.geom('geom1').feature.create('c1', 'Circle');
model.geom('geom1').feature.create('r1', 'Rectangle');
model.geom('geom1').feature.create('arr1', 'Array');
model.geom('geom1').feature.create('r2', 'Rectangle');
model.geom('geom1').feature.create('dif1', 'Difference');
model.geom('geom1').feature.create('r3', 'Rectangle');
model.geom('geom1').feature.create('r4', 'Rectangle');
model.geom('geom1').feature.create('r5', 'Rectangle');
model.geom('geom1').feature.create('dif2', 'Difference');
model.geom('geom1').feature('c1').set('r', '0.5');
model.geom('geom1').feature('r1').set('base', 'center');
model.geom('geom1').feature('r1').set('pos', {'-2.2*wSC' '2e-3'});
model.geom('geom1').feature('r1').set('size', {'wSC' 'tSC'});
model.geom('geom1').feature('arr1').set('size', {'5' '1'});
model.geom('geom1').feature('arr1').set('displ', {'1.1*wSC' '0'});
model.geom('geom1').feature('arr1').selection('input').set({'r1'});
model.geom('geom1').feature('r2').set('base', 'center');
model.geom('geom1').feature('r2').set('pos', {'0' '-0.25'});
model.geom('geom1').feature('r2').set('size', {'1' '0.5'});
model.geom('geom1').feature('dif1').selection('input').set({'c1'});
model.geom('geom1').feature('dif1').selection('input2').set({'r2'});
model.geom('geom1').feature('r3').set('base', 'center');
model.geom('geom1').feature('r3').set('pos', {'-2.56*wSC' '2.2e-3'});
model.geom('geom1').feature('r3').set('size', {'wSC/3' 'tSC*500'});
model.geom('geom1').feature('r4').set('base', 'center');
model.geom('geom1').feature('r4').set('pos', {'2.56*wSC' '2.2e-3'});
model.geom('geom1').feature('r4').set('size', {'wSC/3' 'tSC*500'});
model.geom('geom1').feature('r5').set('base', 'center');
model.geom('geom1').feature('r5').set('pos', {'0' '1.01e-3'});
model.geom('geom1').feature('r5').set('size', {'5.405*wSC' '2e-3'});
model.geom('geom1').feature('dif2').selection('input').set({'r3' 'r4'});
model.geom('geom1').feature('dif2').selection('input2').set({'r5'});
model.geom('geom1').feature('fin').set('repairtol', '1.0E-8');
model.geom('geom1').run;

model.variable.create('var4');
model.variable('var4').model('mod1');
model.variable('var4').set('rho', 'rhoFM');
model.variable('var4').set('muFM', '1+30600*(1-exp(-((Hm[m/A])/295)^2.5))/(Hm[m/A])^0.81+45*exp(-((Hm
    [m/A])/120)^2.5)');
model.variable('var4').set('mu', 'mu0*muFM');
model.variable('var4').selection.geom('geom1', 2);
model.variable('var4').selection.set([2 8]);
model.variable.create('var1');
model.variable('var1').model('mod1');
model.variable('var1').set('rho', 'Ec/Jc*abs(Jz/Jc)^(n-1)');
model.variable('var1').set('mu', 'mu0');
model.variable('var1').selection.geom('geom1', 2);
model.variable('var1').selection.set([3 4 5 6 7]);
model.variable.create('var2');
model.variable('var2').model('mod1');
model.variable('var2').set('rho', 'rhoAir');
```

```
model.variable('var2').set('mu', 'mu0');
model.variable('var2').selection.geom('geom1', 2);
model.variable('var2').selection.set([1]);
model.variable.create('var3');
model.variable('var3').model('mod1');
model.variable('var3').set('Jz', 'H2x-H1y');
model.variable('var3').set('Ez', 'rho*Jz');
model.variable('var3').set('Hm', '((H1+0.001)^2+(H2+0.001)^2)^0.5');
model.variable('var3').set('Hmax', 'Iext(t)/(2*pi*(x^2+y^2)^0.5)');
model.variable('var3').set('Bm', 'mu*Hm');

model.cpl.create('intop1', 'Integration', 'geom1');
model.cpl.create('intop2', 'Integration', 'geom1');
model.cpl.create('intop3', 'Integration', 'geom1');
model.cpl.create('intop4', 'Integration', 'geom1');
model.cpl.create('intop5', 'Integration', 'geom1');
model.cpl.create('aveop1', 'Average', 'geom1');
model.cpl.create('aveop2', 'Average', 'geom1');
model.cpl.create('aveop3', 'Average', 'geom1');
model.cpl.create('aveop4', 'Average', 'geom1');
model.cpl.create('aveop5', 'Average', 'geom1');
model.cpl('intop1').selection.set([3]);
model.cpl('intop2').selection.set([4]);
model.cpl('intop3').selection.set([5]);
model.cpl('intop4').selection.set([6]);
model.cpl('intop5').selection.set([7]);
model.cpl('aveop1').selection.set([3]);
model.cpl('aveop2').selection.set([4]);
model.cpl('aveop3').selection.set([5]);
model.cpl('aveop4').selection.set([6]);
model.cpl('aveop5').selection.set([7]);

model.physics.create('g', 'GeneralFormPDE', 'geom1');
model.physics('g').field('dimensionless').field('H');
model.physics('g').field('dimensionless').component({'H1' 'H2'});
model.physics('g').feature.create('dir1', 'DirichletBoundary', 1);
model.physics('g').feature('dir1').selection.set([1]);
model.physics('g').feature.create('constr1', 'PointwiseConstraint', 1);
model.physics('g').feature('constr1').selection.set([7 12 16 20 24]);
model.physics('g').feature.create('constr2', 'PointwiseConstraint', 1);
model.physics('g').feature('constr2').selection.set([7 12 16 20 24]);
model.physics('g').feature.create('constr3', 'PointwiseConstraint', 1);
model.physics('g').feature('constr3').selection.set([7 12 16 20 24]);
model.physics('g').feature.create('constr4', 'PointwiseConstraint', 1);
model.physics('g').feature('constr4').selection.set([7 12 16 20 24]);
model.physics('g').feature.create('constr5', 'PointwiseConstraint', 1);
model.physics('g').feature('constr5').selection.set([7 12 16 20 24]);

model.mesh.create('mesh1', 'geom1');
model.mesh('mesh1').feature.create('map1', 'Map');
model.mesh('mesh1').feature.create('ftri1', 'FreeTri');
model.mesh('mesh1').feature('map1').selection.geom('geom1', 2);
model.mesh('mesh1').feature('map1').selection.set([3 4 5 6 7]);
```

```
model.mesh('mesh1').feature('map1').feature.create('dis1', 'Distribution');
model.mesh('mesh1').feature('map1').feature.create('dis2', 'Distribution');
model.mesh('mesh1').feature('map1').feature('dis1').selection.set([8 9 13 14 17 18 21 22 25 26]);
model.mesh('mesh1').feature('map1').feature('dis2').selection.set([7 11 12 15 16 19 20 23 24 30]);
model.mesh('mesh1').feature('ftri1').feature.create('dis1', 'Distribution');
model.mesh('mesh1').feature('ftri1').feature('dis1').selection.set([1]);

model.variable('var4').name('Domain:_Ferromagnet');
model.variable('var1').name('Domain:_Supercond');
model.variable('var2').name('Domain:_Vacuum');
model.variable('var3').name('Entire_Sphere');

model.view('view1').axis.set('xmin', '-0.0110566667193770409');
model.view('view1').axis.set('ymin', '0.0015790574252605438');
model.view('view1').axis.set('xmax', '-0.009940456598997116');
model.view('view1').axis.set('ymax', '0.0024435482919216156');

model.physics('g').prop('ShapeProperty').set('shapeFunctionType', 'shcurl');
model.physics('g').prop('ShapeProperty').set('order', '1');
model.physics('g').prop('Units').set('DependentVariableQuantity', 'magneticfield');
model.physics('g').prop('Units').set('CustomSourceTermUnit', 'V*m^-2');
model.physics('g').feature('gfeq1').set('f', {'-d(mu,t)*H1'; '-d(mu,t)*H2'});
model.physics('g').feature('gfeq1').set('Ga', {'0' 'Ez'; '-Ez' '0'});
model.physics('g').feature('gfeq1').set('da', {'mu'; '0'; '0'; 'mu'});
model.physics('g').feature('dir1').set('r', {'-Hmax*y/(x^2+y^2)^0.5'; '0'});
model.physics('g').feature('dir1').set('useDirichletCondition', {'0'; '1'});
model.physics('g').feature('constr1').set('constraintExpression', 'intop1(Jz)-Iext(t)');
model.physics('g').feature('constr2').set('constraintExpression', 'intop2(Jz)-Iext(t)');
model.physics('g').feature('constr3').set('constraintExpression', 'intop3(Jz)-Iext(t)');
model.physics('g').feature('constr4').set('constraintExpression', 'intop4(Jz)-Iext(t)');
model.physics('g').feature('constr5').set('constraintExpression', 'intop5(Jz)-Iext(t)');

model.mesh('mesh1').feature('size').set('hauto', 6);
model.mesh('mesh1').feature('map1').feature('dis1').set('elemratio', '4');
model.mesh('mesh1').feature('map1').feature('dis1').set('method', 'geometric');
model.mesh('mesh1').feature('map1').feature('dis1').set('elemcount', '60');
model.mesh('mesh1').feature('map1').feature('dis1').set('symmetric', true);
model.mesh('mesh1').feature('map1').feature('dis1').set('type', 'predefined');
model.mesh('mesh1').feature('map1').feature('dis2').set('numelem', '1');
model.mesh('mesh1').feature('ftri1').feature('dis1').set('elemratio', '32');
model.mesh('mesh1').feature('ftri1').feature('dis1').set('method', 'geometric');
model.mesh('mesh1').feature('ftri1').feature('dis1').set('elemcount', '100');
model.mesh('mesh1').feature('ftri1').feature('dis1').set('symmetric', true);
model.mesh('mesh1').feature('ftri1').feature('dis1').set('type', 'predefined');
model.mesh('mesh1').feature('ftri1').feature('dis1').set('reverse', true);
model.mesh('mesh1').run;

model.frame('material1').sorder(1);

model.study.create('std1');
model.study('std1').feature.create('time', 'Transient');

model.sol.create('sol1');
```

```
model.sol('sol1').study('std1');
model.sol('sol1').attach('std1');
model.sol('sol1').feature.create('st1', 'StudyStep');
model.sol('sol1').feature.create('v1', 'Variables');
model.sol('sol1').feature.create('t1', 'Time');

model.study('std1').feature('time').set('initstudyhide', 'on');
model.study('std1').feature('time').set('initsolhide', 'on');
model.study('std1').feature('time').set('notstudyhide', 'on');
model.study('std1').feature('time').set('notsolhide', 'on');

model.result.dataset.create('cln1', 'CutLine2D');
model.result.dataset.create('cln2', 'CutLine2D');
model.result.dataset.create('cln3', 'CutLine2D');
model.result.dataset.create('cln4', 'CutLine2D');
model.result.dataset.create('cln5', 'CutLine2D');
model.result.create('pg1', 'PlotGroup2D');
model.result.create('pg2', 'PlotGroup1D');
model.result.create('pg3', 'PlotGroup1D');
model.result.create('pg4', 'PlotGroup1D');
model.result.create('pg5', 'PlotGroup1D');
model.result('pg1').feature.create('surf1', 'Surface');
model.result('pg2').feature.create('lngr1', 'LineGraph');
model.result('pg2').feature.create('lngr2', 'LineGraph');
model.result('pg2').feature.create('lngr3', 'LineGraph');
model.result('pg2').feature.create('lngr4', 'LineGraph');
model.result('pg2').feature.create('lngr5', 'LineGraph');
model.result('pg3').feature.create('lngr1', 'LineGraph');
model.result('pg4').feature.create('glob1', 'Global');
model.result('pg5').feature.create('glob1', 'Global');
model.result('pg5').feature.create('glob2', 'Global');
model.result('pg5').feature.create('glob3', 'Global');
model.result('pg5').feature.create('glob4', 'Global');
model.result('pg5').feature.create('glob5', 'Global');
model.result.export.create('img1', 'Image2D');
model.result.export.create('img2', 'Image1D');
model.result.export.create('img3', 'Image1D');
model.result.export.create('img4', 'Image1D');
model.result.export.create('plot1', 'Plot');
model.result.export.create('plot2', 'Plot');
model.result.export.create('plot3', 'Plot');

model.study('std1').feature('time').set('tlist', 'range(0,1/tRampDuration,tRampDuration)');

model.sol('sol1').attach('std1');
model.sol('sol1').feature('st1').name('Compile_Equations:_Time_Dependent');
model.sol('sol1').feature('st1').set('studystep', 'time');
model.sol('sol1').feature('v1').set('control', 'time');
model.sol('sol1').feature('v1').feature('mod1_H').name('mod1.H');
model.sol('sol1').feature('t1').set('atolglobal', '1e-7');
model.sol('sol1').feature('t1').set('initialstepbdf', '1e-5');
model.sol('sol1').feature('t1').set('maxstepbdf', '1e-4');
model.sol('sol1').feature('t1').set('atolglobalmethod', 'unscaled');
```

```
model.sol('sol1').feature('t1').set('tlist', 'range(0,1/tRampDuration,tRampDuration)');
model.sol('sol1').feature('t1').set('control', 'time');
model.sol('sol1').feature('t1').set('bwinitstepfrac', '1.0');
model.sol('sol1').feature('t1').feature('dDef').set('mumpsalloc', '4');
model.sol('sol1').feature('t1').feature('dDef').set('errorchk', 'off');
model.sol('sol1').feature('t1').feature('fcDef').set('jtech', 'onevery');
model.sol('sol1').runAll;

model.result.dataset('cln1').name('Superconductor_1');
model.result.dataset('cln1').set('genpoints', {'-2.7*wSC' '2e-3'; '-1.7*wSC' '2e-3'});
model.result.dataset('cln2').name('Superconductor_2');
model.result.dataset('cln2').set('spacevars', {'cln1x'});
model.result.dataset('cln2').set('genpoints', {'-1.6*wSC' '2e-3'; '-.6*wSC' '2e-3'});
model.result.dataset('cln3').name('Superconductor_3');
model.result.dataset('cln3').set('spacevars', {'cln1x'});
model.result.dataset('cln3').set('genpoints', {'-.5*wSC' '2e-3'; '.5*wSC' '2e-3'});
model.result.dataset('cln4').name('Superconductor_4');
model.result.dataset('cln4').set('spacevars', {'cln1x'});
model.result.dataset('cln4').set('genpoints', {'.6*wSC' '2e-3'; '1.6*wSC' '2e-3'});
model.result.dataset('cln5').name('Superconductor_5');
model.result.dataset('cln5').set('spacevars', {'cln1x'});
model.result.dataset('cln5').set('genpoints', {'1.7*wSC' '2e-3'; '2.7*wSC' '2e-3'});
model.result('pg1').set('view', 'view1');
model.result('pg1').feature('surf1').set('descr', '');
model.result('pg1').feature('surf1').set('expr', 'Hm');
model.result('pg2').name('1D_Plot_Jz');
model.result('pg2').set('data', 'none');
model.result('pg2').set('ylabel', 'Jz*tSC_(A/m)');
model.result('pg2').set('xlabel', 'Arc_length');
model.result('pg2').set('ylabelactive', false);
model.result('pg2').set('xlabelactive', false);
model.result('pg2').feature('lngr1').set('data', 'cln1');
model.result('pg2').feature('lngr1').set('expr', 'Jz*tSC');
model.result('pg2').feature('lngr1').set('descr', 'Jz*tSC');
model.result('pg2').feature('lngr1').set('legendmethod', 'manual');
model.result('pg2').feature('lngr1').set('legends', {'SC1'});
model.result('pg2').feature('lngr1').set('looplevelinput', {'all'});
model.result('pg2').feature('lngr1').set('legend', true);
model.result('pg2').feature('lngr2').set('data', 'cln2');
model.result('pg2').feature('lngr2').set('expr', 'Jz*tSC');
model.result('pg2').feature('lngr2').set('descr', 'Jz*tSC');
model.result('pg2').feature('lngr2').set('legendmethod', 'manual');
model.result('pg2').feature('lngr2').set('legends', {'SC2'});
model.result('pg2').feature('lngr2').set('looplevelinput', {'all'});
model.result('pg2').feature('lngr2').set('legend', true);
model.result('pg2').feature('lngr3').set('data', 'cln3');
model.result('pg2').feature('lngr3').set('expr', 'Jz*tSC');
model.result('pg2').feature('lngr3').set('descr', 'Jz*tSC');
model.result('pg2').feature('lngr3').set('legendmethod', 'manual');
model.result('pg2').feature('lngr3').set('legends', {'SC3'});
model.result('pg2').feature('lngr3').set('looplevelinput', {'all'});
model.result('pg2').feature('lngr3').set('legend', true);
model.result('pg2').feature('lngr4').set('data', 'cln4');
```

```
model.result('pg2').feature('lngr4').set('expr', 'Jz*tSC');
model.result('pg2').feature('lngr4').set('descr', 'Jz*tSC');
model.result('pg2').feature('lngr4').set('legendmethod', 'manual');
model.result('pg2').feature('lngr4').set('legends', {'SC4'});
model.result('pg2').feature('lngr4').set('looplevelinput', {'all'});
model.result('pg2').feature('lngr4').set('legend', true);
model.result('pg2').feature('lngr5').set('data', 'cln5');
model.result('pg2').feature('lngr5').set('expr', 'Jz*tSC');
model.result('pg2').feature('lngr5').set('descr', 'Jz*tSC');
model.result('pg2').feature('lngr5').set('legendmethod', 'manual');
model.result('pg2').feature('lngr5').set('legends', {'SC5'});
model.result('pg2').feature('lngr5').set('looplevelinput', {'all'});
model.result('pg2').feature('lngr5').set('legend', true);
model.result('pg3').name('1D_Plot_B');
model.result('pg3').set('data', 'none');
model.result('pg3').set('xlabel', 'x-coordinate_(m)');
model.result('pg3').set('xlabelactive', false);
model.result('pg3').feature('lngr1').set('data', 'cln1');
model.result('pg3').feature('lngr1').set('unit', '');
model.result('pg3').feature('lngr1').set('xdataunit', 'm');
model.result('pg3').feature('lngr1').set('xdata', 'expr');
model.result('pg3').feature('lngr1').set('expr', 'mu0*H2');
model.result('pg3').feature('lngr1').set('xdataexpr', 'x');
model.result('pg3').feature('lngr1').set('descr', '');
model.result('pg3').feature('lngr1').set('xdatadescr', 'x-coordinate');
model.result('pg3').feature('lngr1').set('looplevelinput', {'all'});
model.result('pg3').feature('lngr1').set('unit', '');
model.result('pg4').name('1D_Plot_P');
model.result('pg4').set('ylabel', 'Losses_P_(W/m)');
model.result('pg4').set('title', 'Superconductor_ac-losses_P_(W/m)');
model.result('pg4').set('titletype', 'manual');
model.result('pg4').set('ylabelactive', true);
model.result('pg4').set('xlabel', 'Time_(s)');
model.result('pg4').set('xlabelactive', true);
model.result('pg4').feature('glob1').set('expr', {'intop1(Jz*Ez)' 'intop2(Jz*Ez)' 'intop3(Jz*Ez)' '
    intop4(Jz*Ez)' 'intop5(Jz*Ez)'});
model.result('pg4').feature('glob1').set('unit', {'W/m' 'W/m' 'W/m' 'W/m' 'W/m'});
model.result('pg4').feature('glob1').set('descr', {'' '' '' '' ''});
model.result('pg5').name('1D_Plot_V');
model.result('pg5').set('axislimits', 'on');
model.result('pg5').set('ymin', '0');
model.result('pg5').set('ymax', '4.999999873689376E-6');
model.result('pg5').set('xmin', '1.52587890625E-5');
model.result('pg5').set('xmax', '310');
model.result('pg5').feature('glob1').set('expr', {'aveop1(Ez)'});
model.result('pg5').feature('glob1').set('xdatadescr', 'intop1(Jz)');
model.result('pg5').feature('glob1').set('unit', {'V/m'});
model.result('pg5').feature('glob1').set('xdataexpr', 'intop1(Jz)');
model.result('pg5').feature('glob1').set('xdataunit', 'A');
model.result('pg5').feature('glob1').set('descr', {''});
model.result('pg5').feature('glob1').set('xdata', 'expr');
model.result('pg5').feature('glob2').set('expr', {'aveop2(Ez)'});
model.result('pg5').feature('glob2').set('xdatadescr', 'intop2(Jz)');
```

```
model.result('pg5').feature('glob2').set('unit', {'V/m'});
model.result('pg5').feature('glob2').set('xdataexpr', 'intop2(Jz)');
model.result('pg5').feature('glob2').set('xdataunit', 'A');
model.result('pg5').feature('glob2').set('descr', {'SC2'});
model.result('pg5').feature('glob2').set('xdata', 'expr');
model.result('pg5').feature('glob3').set('expr', {'aveop3(Ez)'});
model.result('pg5').feature('glob3').set('xdatadescr', 'intop3(Jz)');
model.result('pg5').feature('glob3').set('unit', {'V/m'});
model.result('pg5').feature('glob3').set('xdataexpr', 'intop3(Jz)');
model.result('pg5').feature('glob3').set('xdataunit', 'A');
model.result('pg5').feature('glob3').set('descr', {'SC3'});
model.result('pg5').feature('glob3').set('xdata', 'expr');
model.result('pg5').feature('glob4').set('expr', {'aveop4(Ez)'});
model.result('pg5').feature('glob4').set('xdatadescr', 'intop4(Jz)');
model.result('pg5').feature('glob4').set('unit', {'V/m'});
model.result('pg5').feature('glob4').set('xdataexpr', 'intop4(Jz)');
model.result('pg5').feature('glob4').set('xdataunit', 'A');
model.result('pg5').feature('glob4').set('descr', {'SC4'});
model.result('pg5').feature('glob4').set('xdata', 'expr');
model.result('pg5').feature('glob5').set('expr', {'aveop5(Ez)'});
model.result('pg5').feature('glob5').set('xdatadescr', 'intop5(Jz)');
model.result('pg5').feature('glob5').set('unit', {'V/m'});
model.result('pg5').feature('glob5').set('xdataexpr', 'intop5(Jz)');
model.result('pg5').feature('glob5').set('xdataunit', 'A');
model.result('pg5').feature('glob5').set('descr', {'SC5'});
model.result('pg5').feature('glob5').set('xdata', 'expr');
```

A.2 Abbreviation Index

Karlsruher Schriftenreihe zur Supraleitung
(ISSN 1869-1765)

Herausgeber: Prof. Dr.-Ing. M. Noe, Prof. Dr. rer. nat. M. Siegel

**Die Bände sind unter www.ksp.kit.edu als PDF frei verfügbar
oder als Druckausgabe bestellbar.**

Band 001
Christian Schacherer
**Theoretische und experimentelle Untersuchungen zur
Entwicklung supraleitender resistiver Strombegrenzer.** 2009
ISBN 978-3-86644-412-6

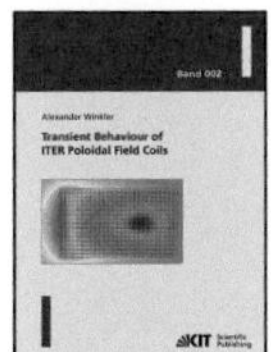

Band 002
Alexander Winkler
Transient behaviour of ITER poloidal field coils. 2011
ISBN 978-3-86644-595-6

Band 003
André Berger
**Entwicklung supraleitender, strombegrenzender
Transformatoren.** 2011
ISBN 978-3-86644-637-3

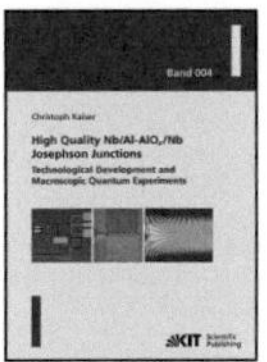

Band 004
Christoph Kaiser
**High quality Nb/Al-AlOx/Nb Josephson junctions. Technological
development and macroscopic quantum experiments.** 2011
ISBN 978-3-86644-651-9

Band 005
Gerd Hammer
**Untersuchung der Eigenschaften von planaren Mikrowellen-
resonatoren für Kinetic-Inductance Detektoren bei 4,2 K.** 2011
ISBN 978-3-86644-715-8

Band 006
Olaf Mäder
Simulationen und Experimente zum Stabilitätsverhalten von HTSL-Bandleitern. 2012
ISBN 978-3-86644-868-1

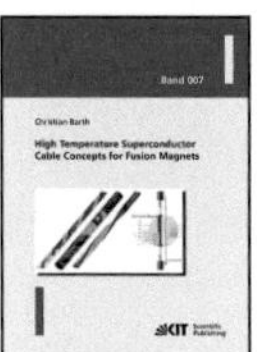

Band 007
Christian Barth
High Temperature Superconductor Cable Concepts for Fusion Magnets. 2013
ISBN 978-3-7315-0065-0

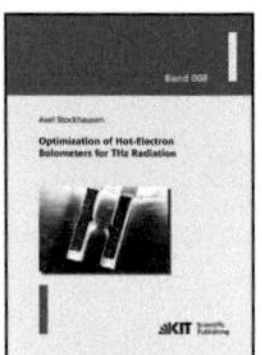

Band 008
Axel Stockhausen
Optimization of Hot-Electron Bolometers for THz Radiation. 2013
ISBN 978-3-7315-0066-7

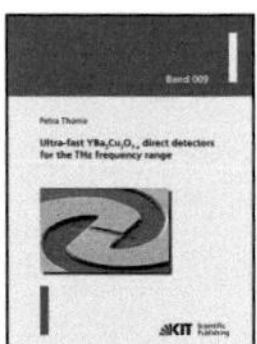

Band 009
Petra Thoma
Ultra-fast YBa$_2$Cu$_3$O$_{7-x}$ direct detectors for the THz frequency range. 2013
ISBN 978-3-7315-0070-4

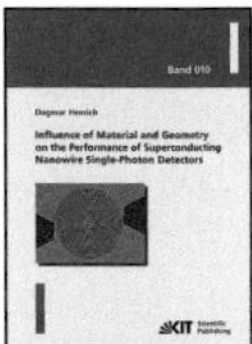

Band 010
Dagmar Henrich
Influence of Material and Geometry on the Performance of Superconducting Nanowire Single-Photon Detectors. 2013
ISBN 978-3-7315-0092-6

Band 011
Alexander Scheuring
Ultrabreitbandige Strahlungseinkopplung in THz-Detektoren. 2013
ISBN 978-3-7315-0102-2

Band 012
Markus Rösch
Development of lumped element kinetic inductance detectors for mm-wave astronomy at the IRAM 30 m telescope. 2013
ISBN 978-3-7315-0110-7

Band 013
Johannes Maximilian Meckbach
Superconducting Multilayer Technology for Josephson Devices. 2013
ISBN 978-3-7315-0122-0

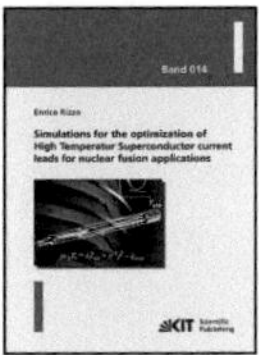

Band 014
Enrico Rizzo
Simulations for the optimization of High Temperature Superconductor current leads for nuclear fusion applications. 2014
ISBN 978-3-7315-0132-9

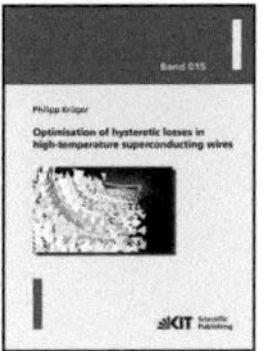

Band 015
Philipp Krüger
Optimisation of hysteretic losses in high-temperature superconducting wires. 2014
ISBN 978-3-7315-0185-5